THE ENABLING ENVIRONMENT FOR DISASTER RISK FINANCING IN THE PHILIPPINES

COUNTRY DIAGNOSTICS ASSESSMENT

DECEMBER 2024

ASIAN DEVELOPMENT BANK

Contents

Boxes

Acknowledgments

This report was prepared under the technical assistance (TA) 6561: Strengthening the Enabling Environment for Disaster Risk Financing (Phase 2). The TA was executed by the Asian Development Bank (ADB) in collaboration with the Government of the Philippines.

Charlotte Benson, principal disaster risk management specialist, Climate Change and Disaster Risk Management Division, Climate Change and Sustainable Development Department, ADB and Arup Chatterjee, principal financial sector specialist, Finance Sector Office, Sectors Group, ADB, provided direction and technical advice for this report.

The report benefited significantly from discussions with and comments from Kelly Bird, Country Director, ADB Philippine Country Office, Southeast Asia Department (SERD); Jose Antonio R. Tan III, Director, Public Management, Financial Sector, and Trade Division (SEPF), SERD; Stephen Schuster, Principal Financial Sector Specialist, SEPF, SERD; Benita Ainabe, Senior Financial Sector Specialist, SEPF, SERD; Aekapol Chogvilaivan, Senior Economist, SEPF, SERD; Ayako Inagaki, Director, Human and Social Development Division (SEHS), SERD; and Eduardo Banzon, Principal Health Specialist, SEHS, SERD.

The report was produced by a team of ADB consultants comprising international consultants Rodolfo Wehrhahn (team leader, insurance and capital market regulatory specialist), Arman Oza (agriculture and microinsurance insurance specialist), Mark O'Donnell (public finance specialist), and Thalia Alice Georgiou (health insurance specialist); national consultant Shayne Rose Bulos (insurance industry specialist); and ADB consultant Maria Cristina Pascual (project coordinator).

The report benefited extensively from kind interaction with many people in key organizations, to whom the team expresses great appreciation for representatives for their time and candid opinions.

Philippine Government Agencies
    Department of Finance
    Department of Budget Management
    Commission of Audit
    Department of Health
    National Disaster Risk Reduction and Management Council
    Office of Civil Defense
    Department of Agriculture
    National Economic and Development Authority
    Climate Change Commission

Philippine Health Insurance Corporation
Department of the Interior and Local Government
Philippine Atmospheric, Geophysical and Astronomical Service Administration
Philippine Institute of Volcanology and Seismology
Department of Social Welfare and Development
Department of Public Works and Highways
Insurance Commission
Government Services Insurance System
Social Security System
Philippine Red Cross
Philippine Crop Insurance Corporation
National Reinsurance Corporation of Philippines
Local Government Unit, Quezon City
Local Government Unit, Municipality of Malabon
Local Government Unit, Municipality of Taal
Local Government Unit, Northern Samar
Local Government Unit, Agusan del Norte Provincial Government

Private Sector
Actuarial Society of Philippines
Malayan Insurance Company, Inc.
Microinsurance Mutual Benefit Association of Philippines Inc.
Philippine Insurers and Reinsurers Association
Philippine Life Insurance Association, Inc.
1 Cooperative Insurance System of Philippines Life and General Insurance
Pioneer Insurance and Surety Corporation
Willis Tower Watson Philippines

Development Partners
Department of Foreign Affairs and Trade, Australia
World Bank

Abbreviations

1CISP	–	1 Cooperative Insurance System of the Philippines
ADB	–	Asian Development Bank
BSP	–	Bangko Sentral ng Pilipinas
CDC	–	Center for Disease Control and Prevention
COVID-19	–	coronavirus disease
DRF	–	disaster risk financing
DRM	–	disaster risk management
GDP	–	gross domestic product
GPS	–	government premium subsidies
GVA	–	gross value-added
HMO	–	health maintenance organization
ICP	–	Insurance Core Principles
IFRS	–	International Financial Reporting Standards
IMF	–	International Monetary Fund
IPSAS	–	International Public Sector Accounting Standards
IRM	–	interim reimbursement mechanism
MBA	–	mutual benefit association
MFI	–	microfinance institution
MI-MBA	–	microinsurance mutual benefit association
PAGASA	–	Philippine Atmospheric, Geophysical and Astronomical Services Administration
PCIC	–	Philippine Crop Insurance Corporation
PCIF	–	Philippine Catastrophe Insurance Facility
PhilHealth	–	Philippine Health Insurance Corporation
PHIVOLCS	–	Philippines Institute of Volcanology and Seismology
PhilCZ	–	Philippine Inter-Agency Committee on Zoonoses
PIDSR	–	Philippine Integrated Disease Surveillance and Response
PIRA	–	Philippine Insurers and Reinsurers Association
PPE	–	personal protective equipment
PSE	–	Philippine Stock Exchange
RBC2	–	Risk-Based Capital Framework 2
RSBSA	–	Registry System for Basic Sectors in Agriculture
SMEs	–	small and medium-sized enterprises
SSS	–	Social Security System
TA	–	technical assistance
UHC	–	universal health care
VAT	–	value-added tax

Currency Unit	**–**	**Philippine Peso (₱)**
₱1.00	–	$48.70
$1.00	–	₱0.0205338809

Executive Summary

The Philippines faces numerous natural hazards that lead to disasters yearly and result in significant financial losses. In addition, the coronavirus disease (COVID-19) pandemic has had a long-lasting severe impact on the economy through medical costs, lockdowns that reduced productivity and created unemployment, and trade and supply chain interruptions that tested financing capacity.

This country diagnostics assessment focuses on disaster risk financing (DRF) and the enabling environment in the Philippines to better understand how the country can improve disaster response. The report efficiently and effectively uses existing financing instruments and introduces additional instruments to enhance the country's financial resilience to disasters, epidemics, and pandemics. The assessment covers risk retention and risk transfer instruments in insurance, reinsurance, and capital market.

An Asian Development Bank–World Bank methodology is applied to assess possible impediments to the effective functioning of self-insurance or risk retention financing instruments of the government. This methodology has been adapted to include pandemic and epidemic risk. The disaster risk retention instrument insights gained from the questionnaire responses are complemented by analyzing existing publicly available information, carrying out discussions with national and local government agencies, and exploiting international best practices.

The modified version of the W&W Development Framework is used to accommodate international good practice and public and private sector stakeholders' inputs. This allows insight into existing or perceived demand and supply barriers restricting the development of an enabling environment for disaster risk transfer instruments. Six areas relevant to the development of insurance and capital market solutions for DRF are reviewed within this framework. These include government policies; social protection policies; unlicensed competition; economic conditions; the credibility of the insurance, reinsurance, and capital market providers; and product appeal.

A risked-layered structure is proposed to stimulate, develop, and implement financially sustainable and scalable DRF strategies and solutions. The assessment makes recommendations to enhance the enabling environment for public sector DRF instruments, insurance, reinsurance, and capital markets solutions. The list of recommendations are presented as follows.

Recommendations for Strengthening the Enabling Environment for Disaster, Epidemic, and Pandemic Risk Financing

Recommendations	Timing and References
Appropriation for the National Disaster Risk Reduction and Management Fund is determined annually, compromising long-term planning.	Near term. Para. 51
Consider automatic appropriation status for National Disaster Risk Reduction and Management Fund funding based on a percentage of gross domestic product or other formula to be determined	
The local special purpose disaster funds (Local Disaster Risk Reduction and Management Fund) have not been fully used in the past.	Immediate. Para. 53
Consider separation of the Local Disaster Risk Reduction and Management Fund for disaster risk reduction and preparedness purposes and post-disaster purposes	
The National Tax Allotment methodology for local government does not recognize the impact of disaster and climate risks, yet climate risk and disaster risk is differentiated across the country.	Immediate. Paras. 57 to 58
Review the Local Disaster Risk Reduction and Management Fund special trust fund, carry forward rules, and consider reviewing the National Tax Allotment methodology for local government to take account of relative climate change and disaster risk.	
While a disaster risk sub-classification is already established within the government's contingent liability classification, a centralized fiscal risk register incorporating disasters is needed.	Near term. Para. 60
Establish a central register of contingent liabilities that identifies and assigns indicative valuations to disaster, epidemic, and pandemic-related fiscal risks.	
Financing for disaster risk management exists in potentially significant budget sums beyond the main program and the National Disaster Risk Reduction and Management Fund in mainstream, regular budgets.	Immediate. Para. 61
Conduct a Disaster Risk Management Public Expenditure and Institutional Review.	
The 2019 Open Budget Survey identified a gap in registration, valuation, and disclosure of disaster-related fiscal risks and contingent liabilities.	Immediate. Para. 62
Investigate the use of temporary expenditure markers to identify the costs of disasters at event level.	
A shortage of physicians and nurses and insufficient supply of resources across the country limit response to pandemics	Near term. Para. 77
Strengthen financing in healthcare infrastructure to ensure adequate health resilience to epidemics and pandemics, especially in remote areas.	
Availability of historic data is insufficient, aggregated and granular, and is necessary to develop and design suitable disaster risk financing instruments.	Near term. Para. 78
Continue improving risk modeling and make the resulting risk data more widely available to government agencies, the insurance sector, and other nonsovereign users.	

continued on next page

Table   *continued*

Recommendations	Timing and References
The inclusion of data from private health facilities is important to ensure comprehensive coverage of all potential points of patient presentation.	Near term. Para. 79
Evolve integrated health surveillance systems through collaborative effort between the government, the World Health Organization, and other parties involved in joint external evaluation.	
While important pandemic and epidemic data collection and monitoring arrangements are in place, some areas require strengthening.	Near term. Para. 79
Make data on the total cost of health delivery available.	
A portal for agriculture, where all data available can be integrated, does not exist.	Near term, Para. 80
Launch a dedicated portal for agriculture, where all available data can be integrated and made accessible to registered government and private sector entities.	
Substantial investments in agritech will be needed so the agriculture sector can withstand disasters efficiently and to enhance its insurability.	Near term, Para. 81
Channel substantial investments in agritech to increase efficiency and predictability, and thus reduce agriculture risk to a level where insurance becomes sustainable and affordable.	
The use of parametric catastrophe risk insurance needs improvement.	Immediate. Paras. 87, 88
Enhance disbursement of claims payments to the ultimate beneficiaries, and further reduce the basis risk.	
PhilHealth's role in financing direct healthcare costs arising from epidemics and pandemics is unclear.	Immediate. Para. 111
Clarify the role and responsibility of PhilHealth in financing direct healthcare costs arising from epidemic and pandemic claims	
Zoonotic disease management is underdeveloped, which is likely to make the risk of economic losses from epidemics and pandemics high.	Near term. Para. 117
Improve the functionality and effectiveness of local government units in zoonosis control and prevention, heighten public health literacy, and enhance public health campaigns.	
The agriculture sector depends heavily on subsidies that, while supporting the use of insurance, also create certain behavioral anomalies disincentivizing individual risk management.	Near term. Para. 135
Use subsidies to encourage and enhance farmers' risk management skills to reduce the cost of insurance.	
The operational independence of the Insurance Commission suffers from certain limitations.	Near term. Para. 140
Improve the operational independence of the Insurance Commission, accompanied by measures to increase its accountability to the government.	
Planned additional increments on already very high minimal capital could reduce competition and increase insurance costs.	Near term. Para. 141
Maintain the minimum capital requirements at the current level of ₱900 million but strengthen monitoring and compliance with the Risk-Based Capital Framework requirements.	

continued on next page

Table   *continued*

Recommendations	Timing and References
Key participants in the provision of insurance to the public—namely, *Philippine Crop Insurance Corporation, the Government Service Insurance System, and the Philippines Health Insurance Corporation*—lag compliance with insurance technical practices and need effective insurance supervision. **Assign the supervision of insurance-related technical matters of Philippine Crop Insurance Corporation, Government Service Insurance System, and PhilHealth to the Insurance Commission.**	Near term. Para. 147
Private sector involvement in agriculture insurance remains very limited despite efforts to encourage it. **Continue encouraging private sector involvement in agriculture insurance, especially for commercial farmers and nontraditional crops.**	Near term. Para. 158
Access to asset and property disaster risk insurance is limited in certain areas of the country. **Provide technically sound catastrophe risk insurance—through the Philippine Catastrophe Insurance Facility, if established—to any part of the country where residential and other buildings are officially allowed to be built.**	Near term. Para. 169
Agriculture involves a wide range of risks with varying frequency and severity, making it necessary to implement risk sharing arrangements, between all stakeholders, that are not in place. **Through smart subsidies, encourage a risk-sharing arrangement between farmers' cooperatives and commercial insurer.**	Near term. Para. 171
Agriculture reinsurance is not readily available at an affordable price. **Establish an agriculture insurance pool to fill the gap between domestic and international reinsurance markets allowing higher domestic agriculture risk retention.**	Near term. Para. 172
Significant business revenue losses remained uninsured during the pandemic. **Consider multiyear business interruption insurance for the government.**	Near term. Para. 184
The issuance and trading of insurance-linked securities, like catastrophe bonds issued for the benefit of the government, remain outside the Philippine Stock Exchange. **Further develop the Philippine Stock Exchange for efficient issuance of insurance-linked securities for financing of disaster risk related losses.**	Near term. Para. 189
Most microinsurance mutual benefit associations are part of microfinance institutions and thus only serve their members. **Develop strategies for microinsurance mutual benefit associations for tapping the low-income market beyond micro-credit clientele.**	Near term. Para. 203
All insurance products face value-added tax and stamp duty. **Exempt all microinsurance products from taxation.**	Near term. Para. 204

Note: "Immediate" is within 1 year; "near term" is 1 to 3 years.
Source: Asian Development Bank.

Introduction

1.1 Background

1.   **Disasters triggered by natural hazards, epidemics, and pandemics delay long-term development and hamper efforts to reduce poverty in developing countries in Asia and the Pacific.** Disasters directly damage or destroy infrastructure and disrupt economic activities and the provision of services. They place countries on lower long-term growth trajectories, push vulnerable communities deeper into poverty, and force adjustments in short- and longer-term development targets and goals. They can place significant fiscal strain on governments, businesses, and individual households, particularly if these are financially ill-prepared. Delays and shortages in available funding can significantly exacerbate the consequences of direct physical losses, extending the time to rebuild. Government officials, policymakers, and insurance regulators from developing countries across Asia and the Pacific are therefore seeking to strengthen their financial preparedness for disasters, epidemics, and pandemics, smoothing the cost of realized events over time and ensuring timely availability of post-disaster funding. A strong enabling environment for disaster, epidemic, and pandemic risk financing, including for the stimulation of commercial risk transfer markets, is a priority prerequisite for achieving this. This financing is collectively referred to as disaster risk financing (DRF) in this report.

2.   **The severity and long-lasting duration of the coronavirus disease (COVID-19) pandemic tested the financial resilience of countries across the globe.** Box 1 highlights the economic impacts of the COVID-19 pandemic and the unprecedented financing challenges it created in addressing health needs, providing livelihoods and business relief support, and maintaining economic stability.

3.   **The different nature of the financial impact of disasters triggered by natural hazards and long-lasting epidemics and pandemics requires differing financial instruments and enabling measures.** Figure 1 illustrates the timelines and impacted areas of loss impacted by disasters and epidemics and pandemics.

4.   **Enhanced financial preparedness for disasters is an Asian Development Bank (ADB) priority.** The ADB technical assistance (TA) project, *Strengthening the Enabling Environment for Disaster Risk Financing (Phase 2)* under which this report is prepared, is consistent with ADB's 2021 Disaster and Emergency Assistance Policy. This explicitly supports enhanced financing arrangements for disasters, epidemics, and pandemics (ADB 2020d). It is also consistent with the Financial Sector Directional Guide under development by ADB, which calls for building capabilities in emerging and innovative finance areas such as DRF.

Box 1: Impact of the Coronavirus Disease Pandemic

The vast scale of interventions needed to confront the health and economic consequences of the coronavirus disease (COVID-19) pandemic challenged government capacities to manage resources effectively and equitably in unprecedented ways. By the end of 2020, governments had already mobilized $14 trillion in fiscal policy responses of different types. These included additional spending measures, tax relief programs, and loans and loan guarantees—all aimed at funding necessary health services, addressing income losses and keeping economies afloat. Fiscal responses differed across countries—and were much larger in richer countries—but everywhere they represented a very significant departure from normal fiscal policy.[a]

The pandemic was the most devastating shock to the global economy since the Second World War; policies to contain the virus deeply undercut economic activity. The unique character of the recession it sparked poses unfamiliar policy challenges. On the demand side, lockdowns and social distancing measures triggered a sudden stop in spending and left it highly insensitive to policy stimulus. On the supply side, containment measures directly hindered production, with repercussions spreading through local and global supply chains. The overall damage could leave permanent scars if persistent unemployment and bankruptcies follow (BIS 2020).[b]

Across Asia and the Pacific, COVID-19 hit poor and vulnerable people hard, particularly informal workers. Estimates suggest that in the second quarter of 2020, working hours dropped 13.5% in the region, equivalent to 235 million full-time jobs.[c] In the Philippines, a survey by Zero Extreme Poverty Philippines 2030 and the United Nations Development Programme reveals that the pandemic increased the number of households in income poverty, with 83% of respondents experiencing a decline in household income and 34% reporting a complete loss of income. Informal workers suffered the worst: 42% of informal or temporary workers lost their source of income compared to 35% of permanent employees. About 85,000 overseas workers returned to the Philippines and another 300,000 lost their jobs. This resulted in a large loss of remittance income to recipient households.[d] More positively, the unemployment rate showed signs of recovery in 2022, standing at 6.0% in May 2022, down from 7.7% a year earlier.

The COVID-19 pandemic dislocation exposed massive protection gaps in the area of business continuity risk across the world. Less than 1% of the estimated $4.5 trillion global pandemic-induced gross domestic product loss for 2020 is likely to have been covered, reflecting pre-COVID-19 coverage exclusions and restrictions as well as the niche character of business interruption insurance which accounts for less than 2% of the world's property and casualty insurance market.[e]

[a] International Budget Partnership. 2021. *Managing Covid Funds: The Accountability Gap.* https://internationalbudget.org/covid/wpcontent/uploads/2021/05/Report_ English-2.pdf.

[b] Bank for International Settlements. 2020. *Annual Report Economic Report 2020.* https://www.bis.org/publ/arpdf/ar2020e1.pdf.

[c] Asian Development Bank. 2021a. *COVID-19 and the Finance Sector in Asia and the Pacific-Guidance Note.* https://www.adb.org/sites/default/files/institutional-document/761946/covid-19- finance-sector-asia-pacific-guidance-note.pdf.

[d] Food and Agricultural Organization (FAO). 2022. COVID-19 Pandemic Impacts on Asia and the Pacific. https://www.fao.org/3/cb8594en/cb8594en.pdf.

[e] The Geneva Association. 2021. Public-Private Solutions to Pandemic Risk: Opportunities, Challenges and Trade-offs. https://www.unisdr.org/preventionweb/files/77863_publicprivate solutionstopandemicris.pdf.

Source: Authors.

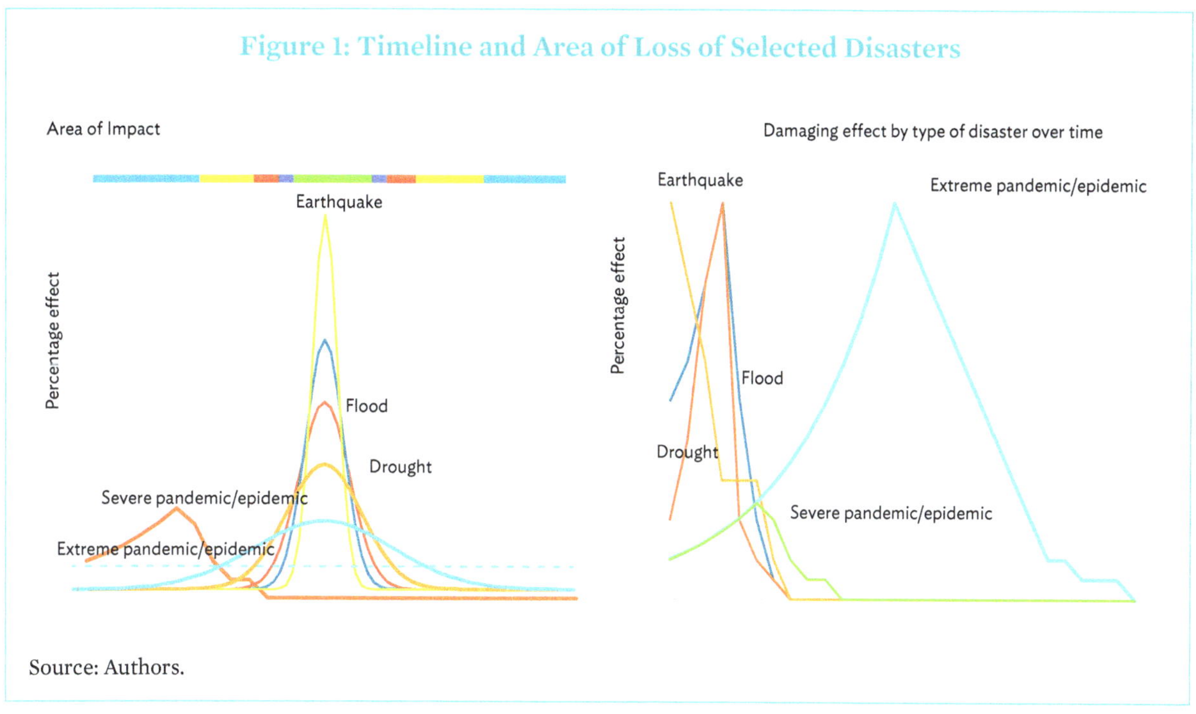

Source: Authors.

5.   **ADB's holistic approach to DRF is reflected in this TA.** ADB strongly advocates an integrated approach to disaster risk management (DRM), seeking to strengthen disaster resilience both through disaster risk reduction and the enhanced management of residual risk. ADB is seeking to enhance financial preparedness for disasters as part of broader efforts to strengthen disaster resilience. It is doing so in close coordination with governments and global and regional DRF initiatives.[1] Likewise, it is closely coordinating with standard-setting bodies (International Association of Insurance Supervisors, International Organization of Securities Commissions, the Basel Committee on Banking Supervision, the Islamic Financial Services Board, the Financial Stability Institute), and the insurance industry. Disaster risk reduction efforts should be the first consideration in addressing disaster risk, tackling the root causes of the issue. DRF solutions should also conform to international financial standards and be designed around the context of broader disaster resilience, financial stability, and financial inclusion, incorporating incentives for disaster risk reduction. This approach should lead to the development and implementation of financially sustainable, scalable DRF strategies and solutions. ADB applies a risk-layered approach to support the appropriate selection of DRM options, including DRF instruments (section 1.2).

6.   **This country diagnostic assessment identifies areas of improvement to promote an enhanced enabling environment for DRF in the Philippines.** Notwithstanding the importance of having and using DRF instruments, these instruments can only be fully effective under certain, often neglected conditions. Assessing and identifying barriers to be removed to create an enabling environment for increased uptake of these instruments is

---

[1] These include the Vulnerable Twenty Group; the Disaster Risk Financing and Insurance Program of the World Bank, Market Global Practice and Global Facility for Disaster Reduction and Recovery; the Pacific Disaster Risk Financing and Insurance Program; and the Asia-Pacific Economic Cooperation and the Organisation for Economic Co-operation and Development promoting the G20/OECD Methodological Framework for Disaster Risk Assessment and Risk Financing.

critical. This country diagnostic is intended to facilitate the development and implementation of appropriate instruments for different layers of risk. It identifies areas of improvement to enhance the enabling environment for public sector DRF solutions as well as insurance, reinsurance, and capital market (IRCM) solutions.

7.   **Recommendations based on the assessment are comprehensively presented in the corresponding sections.** The recommended activities and measures to enhance the enabling environment for key public sector DRF instruments, as well as IRCM solutions, are presented at the end of the corresponding section of relevance. A summary of the main recommendations are listed in the executive summary.

1.2   Risk Layering Approach

8.   **Resilience begins with risk reduction, that is, acting to reduce levels of loss in the event of natural hazards, epidemics, and pandemics.** However, risk cannot be eliminated, so investments in financial preparedness need to be enhanced as well, seeking to ensure that sufficient financing is available to support timely relief, early recovery, and reconstruction efforts.

9.   **Governments can draw on an array of instruments to support enhanced financial preparedness.** These instruments are ideally applied using a risk-layering approach, breaking risk down according to the frequency of occurrence of different types of hazard events, epidemics, and pandemics of varying severity and associated levels of loss and designing bundles of instruments targeting differentiated layers of risk (ADB 2014). Governments should seek to select the most appropriate instruments for each layer of risk based on a range of factors, including the scale of funding needed, the speed with which disbursement is required, and the relative cost-effectiveness of alternative instruments for specific layers of risk. Note that due to the unexpected severity of the COVID-19 pandemic, most of the financing instruments used were *ex-post*. In addition, regulatory forbearance was necessary to support the financial sector (BSP 2021).

10.   **DRF instruments for residual risk begin with risk retention instruments for more frequent, less damaging events (Figure 2).** These include annual contingency budget allocations, reserves, and contingent grant and credit arrangements, all of which are put in place before an event strikes. After an event, governments can also re-allocate budgets, increase borrowing, and raise taxes to provide additional resources.

11.   **Market-based risk transfer solutions provide more cost-efficient financing for medium-level risks, generating higher levels of loss but less frequently.** These include insurance and insurance-linked securities, such as catastrophe (CAT) bonds, and are taken out in anticipation of potential disasters, epidemics, or pandemics. Following major events, governments also appeal to the international community for assistance.

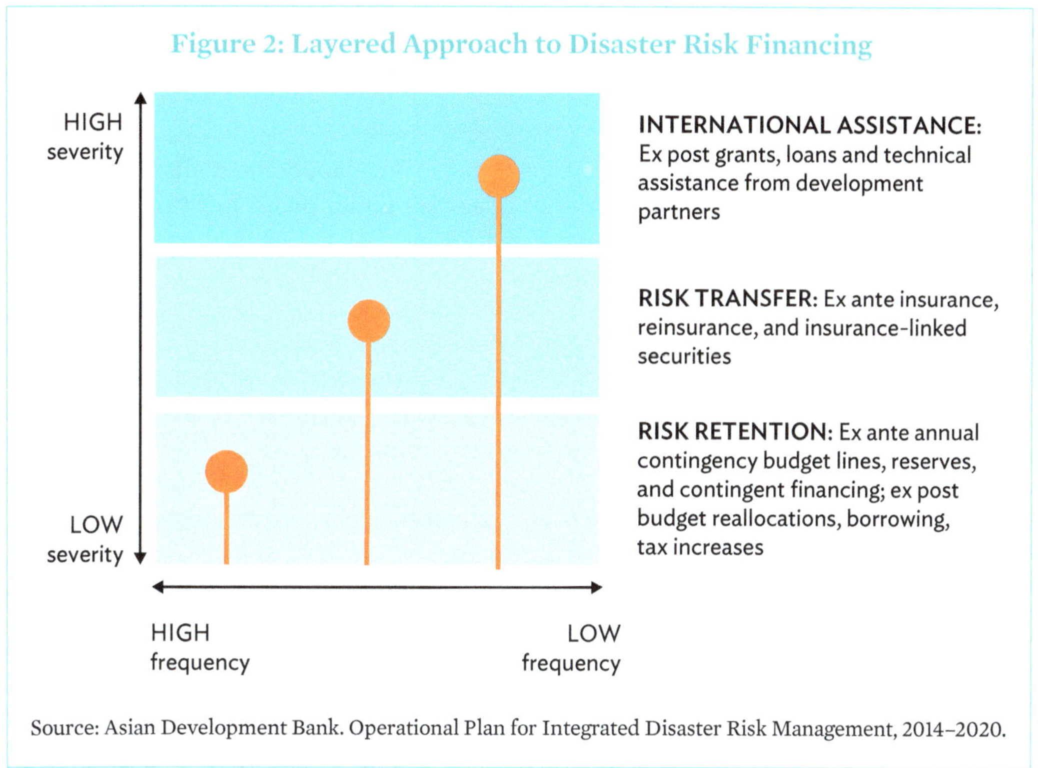

Source: Asian Development Bank. Operational Plan for Integrated Disaster Risk Management, 2014–2020.

12.   **DRF is not only a government responsibility; the private sector and individuals should be encouraged and enabled to share in this.** A similar risk-layering approach is applicable. Decisions on risk reduction, retention, and transfer should be made within the structure of this broader framework, selecting appropriate instruments for each layer of risk. The insurance sector is called to play an important role in this by developing tailor-made products suitable to the Philippines.

13.   **The availability and assortment of instruments selected for a DRF strategy depend on a range of factors.** The most appropriate bundle of instruments depends on the scale of resources required at each layer of loss relative to the scale of resources each instrument can facilitate access to; the speed with which funds are required relative to the disbursement speed of each instrument (Figure 2); the marginal cost of each instrument; individual country circumstances, including prevailing macroeconomic circumstances; the scale of potential events relative to gross domestic product (GDP); government economic, fiscal, and monetary goals and objectives; access to international finance markets; and the market-based cost of borrowing (ADB 2013). For example, if probable maximum losses from extreme events are low relative to GDP, then a country is better able to retain risk. A country with low indebtedness can rely more on post-event borrowing than one with a higher level. The effectiveness of risk transfer instruments also depends crucially on the availability of well-developed and sound domestic insurance and capital market sectors. Among other issues, cultural and religious dimensions are important, and, notably, government policy could potentially crowd out the private insurance sector.

1.3   Diagnostics Methodology

1.3.1   Diagnostic Tool

14.   **The country diagnostics has been undertaken applying a diagnostics tool developed and used to assess four previous countries under phase 1 of this TA project.** However, the tool has been enhanced to cover assessment of the enabling environment for pandemic and epidemic risk financing instruments.

15.   **The diagnostics tool covers both sovereign and nonsovereign risk financing instruments.** Governments apply risk financing instruments to protect their budget. Governments can also play an important role in providing an adequate enabling environment to encourage the supply and uptake of nonsovereign insurance, such as homeowner and commercial property insurance, business interruption cover, health insurance, and crop insurance. In the process, they can reduce the contingent liability falling on government.

16.   **The diagnostics tool generates an overview of current policies and mechanisms for DRF.** It identifies enabling conditions for effective use of well-established DRF instruments and the introduction of new instruments and related barriers and gaps; sets policy priorities for implementing reforms and introducing new DRF instruments; and provides the basis for new or deeper engagement on DRF by governments, regulators, and development partners as part of broader disaster, epidemic and pandemic risk management and public financial management dialogue. The findings of the diagnostic can feed directly into the development of strategies to enhance the financial risk management of disasters, pandemics, and epidemics.

17.   **The tool includes two structured questionnaires.** The first helps assess the framework used by the government to finance disasters, epidemics, and pandemics within its budget, i.e., when disaster risk is retained. The second provides insights into risk transfer instruments. The two questionnaires are critical for evaluating the existing enabling environment for both risk retention and risk transfer instruments.

18.   **The assessment of the enabling environment for the effective use of risk retention instruments is based on a joint ADB and World Bank (2017) questionnaire.** The questionnaire has been enhanced to include epidemic and pandemic risk financing (Box 2).

19.   **The assessment of the enabling environment for the effective use of disaster risk transfer instruments is based on a modified version of the "W&W Development Framework."[2]** This framework was refined to provide a methodology for assessing the DRF landscape and its enabling environment. It focuses on six areas of relevance for the development of disaster, epidemic, and pandemic IRCM solutions:

- **Economic conditions and other support functions** like the disposable budget for insurance, the level of indebtedness as well as data availability, financial sector specialized courts, health facilities, doctors, vaccines and epidemic and pandemic research institutes, risk managers, insurance and other relevant professionals

---

[2]   The W&W Framework has been used on several occasions by Rodolfo Wehrhahn, one of the assessors, to determine barriers to an enabling environment in work done for ADB, the International Monetary Fund, and the World Bank. Relevant areas for an enabling environment as determined in this framework follow from Wehrhahn (2010).

Box 2: Examining the Full Sovereign Disaster Risk Financing Landscape

The Asian Development Bank–World Bank Disaster Risk Finance diagnostic assesses sovereign financial protection against disasters. It has been extended in its current form to include questions for ministries of health and of finance. It includes questions pertaining to risk transfer sovereign arrangements and focuses strongly on risk retention mechanisms. Questions cover the following issues:

- **Assessment of fiscal shocks associated with disasters, epidemics, and pandemics**
  - » contingent liability of the government
  - » fiscal risk assessment of disaster, epidemic, and pandemic shocks
  - » public disclosure of fiscal exposure to disasters, epidemics, and pandemics
- **Ex ante risk financing instruments**
  - » annual contingency budget
  - » dedicated budget lines for risk reduction
  - » dedicated reserve funds
  - » line agency funding
  - » contingent credit and grant arrangements
  - » insurance of public assets
  - » other forms of sovereign insurance
  - » risk transfer arrangements through capital markets
- **Ex post risk financing instruments**
  - » post-event budget reallocations
  - » external assistance
  - » tax increments
  - » government borrowing

Source: Adapted from ADB and the World Bank. 2017. *Assessing Financial Protection against Disasters: A Guidance Note on Conducting a Disaster Risk Finance Diagnostic.*

(e.g., actuaries, adjustors, accountants) necessary for the well-functioning of the providers of IRCM.

- **Government policy on the development of risk transfer instruments,** including the introduction of mandatory insurance protection, risk-pooling structures, and insurance-linked securities;[3] pertinent regulations; and the creation of a level playing field for IRCM activities.
- **Credibility of the private sector offering risk transfer solutions** covering aspects such as the regulatory environment, the solvency of risk carriers, the reputation of insurance and capital markets, as well as the professionality of distribution channels, loss adjusters, brokers, and the availability of infrastructure (e.g., stock exchanges, payment systems, etc.).

---

[3] Insurance-linked securities bonds, including catastrophe bonds and other risk-linked securitization and represented assets whose value is largely driven by the occurrence of events not correlated to the financial markets, allowing for a high degree of diversification. With an insurance linked securities bond the investor is exposed to a well-defined catastrophic or insurable event in addition to the credit risk of the issuer. For this additional exposure, investors are compensated with higher coupons, but if no covered event occurs during the risk period the bonds are redeemed at 100% of face value. When a covered event meets the thresholds in the risk transfer contract, investors stand to lose coupon payments and/or a percentage of the principal. The redemption price of the bonds is reduced accordingly. For more details, see the ADB report entitled *Toolkit for Insurance, Reinsurance, and Capital Market Solutions for Disaster Risk Financing* (revised publication 2022) (forthcoming).

- **Disaster and epidemic and pandemic risk product attractiveness, availability and affordability** including products for governments, corporates, small and medium enterprises, farmers, individual households, and low-income population.
- **Social protection policy,** recognizing that low-income populations should enjoy social protection and at least basic health protection or support in obtaining insurance coverage while insurance solutions for people that can afford the premiums should not be crowded out, and exploring the degree to which social protection complements or might be crowding out market-based solutions.
- **Unlicensed competition,** recognizing that the resilience of insurance providers depends on an adequate level of oversight. Unlicensed entities or those poorly supervised by agencies with insufficient insurance supervision and risk management expertise can destroy the image of insurance and leave their consumers without protection should a disaster cause their insolvency.

20.   **A fuller description of the tool, including the two questionnaires, is presented in the publication *Toolkit for Insurance, Reinsurance, and Capital Market Solutions for Disaster Risk Financing* (revised edition, forthcoming).** The report also presents a generic toolkit for disaster and epidemic and pandemic IRCM solutions focusing on actions to strengthen the enabling environment to support potential DRF instruments and including a glossary of technical terms.

1.3.2   Application of the Diagnostics Tool

21.   **The diagnostics tool is used to determine and confirm DRF practices and gain insight into existing or perceived barriers hindering the development of DRF instruments.** The diagnostics tool is applied through a combination of desk-based work, stakeholder questionnaires, interviews, and group discussions. This wide-ranging approach is taken to accommodate international good practice of countries with successful results and draw in expert judgement on the actions needed to better enable effective use of DRF mechanisms.

22.   **The basic steps are as follows:**

- As a starting point, background information on the DRF strategy of the focus country is gathered. The information is drawn from extensive publications, government documents and websites, insurance and reinsurance industry documents, and capital market analyses.
- The background information is then complemented with extensive questionnaires with open questions on areas relevant to the DRF approach and instruments used in the country. These questionnaires, integral to the diagnostic tool, are sent to relevant stakeholders for their inputs. The insights gained are critical for a robust assessment, and, as such, questions to the stakeholders are explained carefully, stressing the importance of providing comprehensive and open answers.
- Onsite interviews take place with selected stakeholders from both the public sector and IRCM stakeholders, including actuaries, rating agencies, brokers, and auditing firms. These interviews enhance and complete the information gathered through the desk analysis and the questionnaire responses.
- The comprehensive information is then analyzed and gaps between international good practice and current practices identified.

- The recommended actions are discussed with the stakeholders and the feasibility and relevance confirmed before the country diagnostic is finalized.
- Implementation of the recommendations should follow.

23.   **Stakeholders are not necessarily expected to respond to all questions.** Experience shows that the questionnaire will provide wide-ranging responses, including contradictory statements, and leave some questions unanswered. The assessors review and filter the information to draw preliminary conclusions. These conclusions are then verified with stakeholders repeatedly before finalizing the findings and recommendations.

1.3.3   Presentation of the Diagnostic Results

24.   **The country diagnostic reports begin by presenting findings on the broad public sector DRF landscape, including related recommendations.** The results of the diagnostic analysis are then presented and finally summarized in a spider diagram depicting country scoring for each of the six areas of relevance for the development of a strong enabling environment for disaster and epidemic and pandemic IRCM solutions (Figure 3). For each area, an ideal, a realistic, and the current state of the environment are depicted.

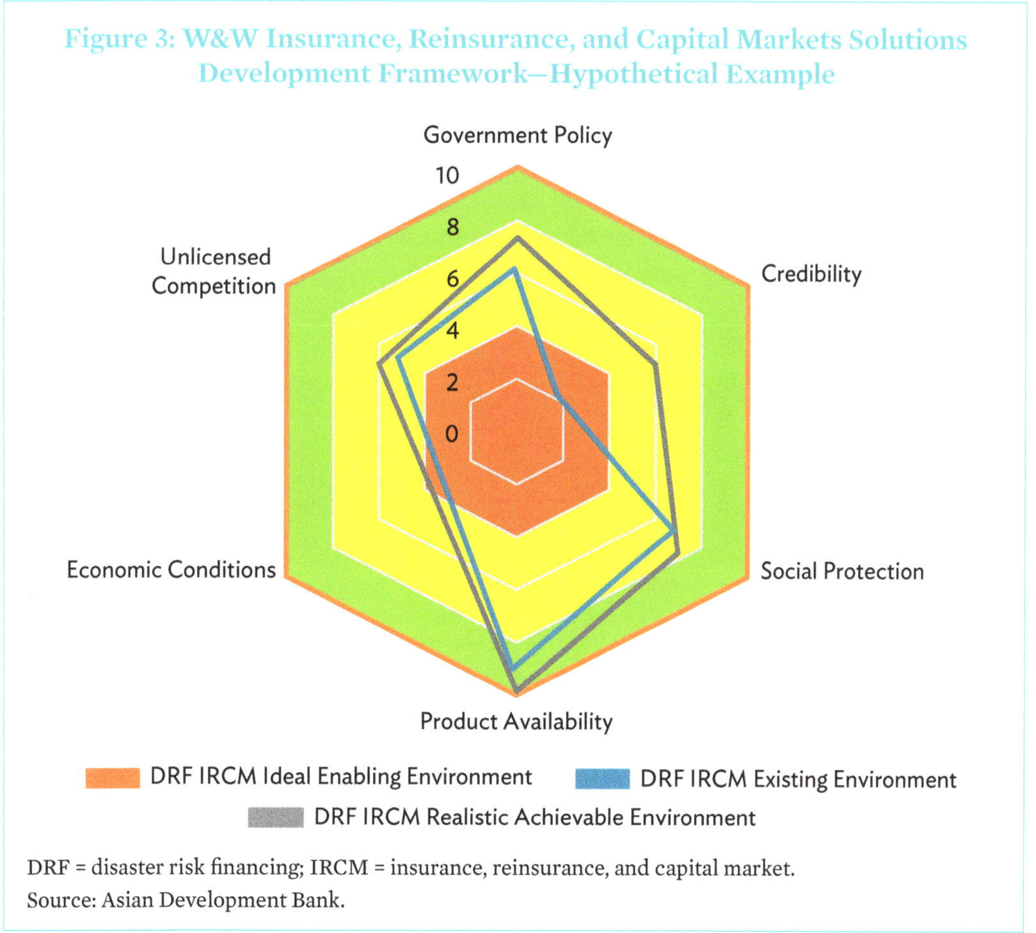

Figure 3: W&W Insurance, Reinsurance, and Capital Markets Solutions Development Framework—Hypothetical Example

DRF = disaster risk financing; IRCM = insurance, reinsurance, and capital market.
Source: Asian Development Bank.

25.   **The ideal enabling conditions for the development of IRCM solutions for each of the six areas are defined.** The assessors define this environment based on international good practice and expert judgement. This step considers the political, cultural, and religious contexts of the marketplace as well as international best practice.

26.   **A reality check defines the next-best IRCM enabling environment that can be achieved.** The ideal enabling environment may be very difficult to achieve in practice, so a realistic enabling environment for each of the six areas is also determined country by country. This is the best achievable enabling environment and is developed drawing on local expertise gained through extensive stakeholder consultation and analysis of the questionnaires to identify likely impediments to achievement of the ideal enabling environment. The ideal and realistic enabling environments will not necessarily differ significantly, resulting in that case in their overlap.

27.   **The current environment is then populated.** Using local expertise and comments from relevant national stakeholders, including government authorities, private sector providers, and professional bodies, the current environment for each axis is determined.

28.   **The methodology used thus depicts the gaps between the current enabling environment for disaster "IRCM solutions" and the ideal and realistic alternatives.** The comparison enables ready identification of areas for action, leading to the development of a strategy and road map for the removal of the gaps. The prioritization of actions to address the gaps should be based on the scale of the gap and reflect reasonable timeframes for completion. Urgent actions are recommended to strengthen the enabling environment in areas of relevance achieving scores of four or below (color red), medium-term actions are recommended for scores between four and six (yellow), and no immediate actions are required for higher scores (green). Where the realistic enabling environment differs from the ideal scenario, that difference between the existing and the realistic scenario rather than between the existing and the ideal scenario is considered when determining the urgency of the needed actions. The absolute scores have no further meaning and should not be used for cross-country comparisons.

The Public Sector Disaster Risk Financing Landscape

2.1   National Budget Resources Context

29.   **The Philippines has a well-developed and advanced funding scheme for disasters.** The country is highly prone to natural hazards and experiences disasters every year. As such, it is not surprising to find an advanced framework in the public sector to cope financially with those events.

30.   **The government developed the National DRF Strategy in 2015. Its first two objectives are to maintain sound fiscal health and develop sustainable financing mechanisms.** The National Disaster Risk Reduction and Management Fund (NDRRMF, also known as the Calamity Fund) is the established national government mechanism to finance the National Disaster Risk Reduction and Management Program (para. 39). Annual appropriation to the Calamity Fund was broadly maintained at pre-COVID-19 levels of around ₱20 billion in 2021 and 2022 (Table 1). Allocations over and above ₱20 billion in 2018 and 2020 were for insurance premiums reallocated rather than spent.

Table 1: National Disaster Risk Reduction and Management Fund Appropriations, 2018–2022

Pesos (₱, billion)	2022	2021	2020	2019	2018
NDRRMP	19.0	13.0	7.5	13.5	7.6
NDRRMP augmentation	0	0	6.8	3.0	2.9
Marawi program	1.0	5.0	3.5	3.5	10.0
CARED	0	0	5.0	0	0
Insurance premiums	0	2.0	0	0	2.0
Total Appropriated (General Appropriation Act)	20.0	20.0	22.8	20.0	22.5

CARED = Comprehensive Aid to Repair Earthquake Damage, NDRRMP = National Disaster Risk Reduction and Management Program.
Source: Government of the Philippines, Department of Budget and Management.

31.   **This appropriation has been maintained in recent years despite challenging circumstances.** Government debt levels and the budget deficit have increased significantly in recent years (Table 2). The Government of Philippines cash budget deficit as a percentage of GDP has also increased (Table 3).

Table 2: National Government Outstanding Debt Stock
(₱ million)

Year	2016	2017	2018	2019	2020	2021
Value	6,603,935	7,130,543	7,780,086	8,220,036	10,253,354	12,152,466

Source: Government of the Philippines, Bureau of Treasury. National Government Outstanding Debt 2016–2021.

Table 3: National Cash Budget Deficit
(% of GDP)

Year	2016	2017	2018	2019	2020	2021
%	2.3%	2.1%	3.1%	3.4%	7.6%	8.6%

Source: Government of the Philippines, Department of Finance.

32.   **The 2022 People's Budget indicated a planned reduction in the deficit to pre-COVID-19 pandemic levels.**[4] In 2021 the government planned to raise the budget deficit to 8.5% of GDP (about ₱1,758.0 billion) to increase spending on COVID-19 pandemic programs and infrastructure projects to help the economy recover. In the 2022 People's Budget (Table 4), the deficit was forecast to decrease to ₱1,665.1 billion, or 7.5% of GDP (Department of Budget and Management 2022, 59). Despite the deficit, the government planned to keep debt around the government's debt threshold of 60% of GDP by 2022 (Department of Budget and Management 2022, p. 38).

33.   **Further, the International Monetary Fund Philippines country report of August 2021 indicated that** *"strong fundamentals and prudent macroeconomic policies have helped to maintain macro-financial stability"* (IMF 2021). In addition to the impact of COVID-19 pandemic, the redistributive effect of the 2018 Mandanas Ruling,[5] which came into effect in 2022, means, other things equal, a redistribution of resources from national to local government with local government forecast to receive an increase in resources of 37.9% (Department of Budget and Management 2022, 6). Despite fiscal pressures on the national government from several sources, including the Mandanas Ruling, COVID-19 pandemic, reduced revenues in 2020 and consequent increases in borrowing, Calamity Fund appropriation has been maintained at ₱20 billion in 2022, which suggests a level of commitment and sustainability as this was achieved within a fiscally sound framework.

Table 4: Philippines Government Planned Borrowing in 2021 and 2022

Borrowing Classification	2021 Budget Proposal (₱, billion)[a]	2022 Budget Proposal (₱, billion)[b]
Net External Financing[c]	442.3	560.6
Net Domestic Financing	2,582.8	1,912.2
Total Planned 2021	**3,025.2**	**2,472.8**

[a] Department of Budget and Management. 2021. People's Proposed Budget, p. 41.
[b] Department of Budget and Management. 2022. People's Proposed Budget, p. 62.
[c] Department of Budget and Management. 2021. People's Proposed Budget, p. 41.
  Note that typographical error resulting in arithmetical issue—bonds, etc., should be ₱286.0 not ₱268.0.
Source: Government of the Philippines, Department of Budget and Management 2021 and 2022.

---

[4] The People's Budget is the Executive's Annual Budget Proposal.
[5] Mandanas et al. v. Executive Secretary et al. (G.R. No. 199802 and G.R. No. 208488) "Mandanas."

2.2 Disaster Risk Financing Overview

34.   **The country's significant disaster risk is acknowledged in recent government reports, including the annual Fiscal Risk Statements.** The 2022 Fiscal Risk Statement report states:[6]

> *"The Philippines is an archipelago that is highly vulnerable to natural disasters and climate risks. Results of the World Risk Index 2020 show that the Philippines was among the 15 countries with the highest risk, and also among the nine countries out of the 15 that has the highest exposure. The country is among the five countries in the Association of Southeast Asian Nations (ASEAN) listed in the highest risk category. Sea-level rise, storms, and earthquakes are among the factors that contribute to the risk profile of these states. (WorldRiskReport 2020).*
>
> *Based on the Global Climate Risk Index 2021 published by German Watch, the Philippines ranked fourth among the countries most affected by extreme weather events in the twenty-year period between 2000 and 2019, after Puerto Rico, Myanmar, and Haiti (Eckstein, Künzel, Schäfer 2021). The country is particularly regularly exposed to tropical cyclones due to its geographical location."* Department of Budget and Management Fiscal Risk Statement 2022 (paragraphs 155 and 156) (DBCC 2022, 64).[7]

The Philippines Statistics Authority reported in October 2020 that *"The damages incurred due to natural extreme events and disasters amounted to ₱463 billion from 2010 to 2019. Agriculture posted the largest share with 62.7% or ₱290 billion followed by infrastructure, and private/ communications with 23.0% or ₱106 billion and 14.3% or ₱66 billion respectively"* (PSA 2020).

In 2019 the Department of Finance (DOF n.d.) indicated that the Philippines is expected to incur average annual losses of ₱177 billion per year of public and private assets due to typhoons and earthquakes. The same report indicated a 40% chance of experiencing loss exceeding ₱989 billion and a 20% chance of experiencing loss exceeding ₱1,525 billion.

35.   **The Philippines has put in place interventions to manage disaster risk.** At the core of these interventions is the National Disaster Risk Reduction and Management Plan (2020–2030) (Office of Civil Defense 2020), established to systematically integrate the various disaster risk management activities, coordination, and financing mechanisms of the government. One of the principles of the plan is a whole of government approach to DRM (Office of Civil Defense 2020).[8] DRM resources are allocated by general appropriation[9] to a series of special purpose funds and directly mandated to national government institutions such as the Office of Civil Defense. Thus, the funding of disaster response uses special purpose funds as well as additional disaster-relevant appropriations. The National Disaster Risk Financing Strategy prioritizes risk transfer and the development of contingent financing

---

[6] Annual Fiscal Risk Statements from 2013 to 2022 can be accessed at Department of Budget and Management (DBM). Fiscal Risks Statement. dbm.gov.ph.

[7] DBCC. 2022. Fiscal-Risks-Statement-2022-for-Circulation.pdf (dbm.gov.ph).

[8] See NDRRMP-Publication-v5 (Revised) 4–22 (ndrrmc.gov.ph), p. 14 (from the paragraph Target Users and Stakeholders).

[9] Technical Note: The People's Budget (the annual budget proposal) identifies both programmed and unprogrammed appropriations. Unprogrammed appropriations are those not yet supported by corresponding resources but are included by Congress in the General Appropriations Act. These are called standby appropriations which authorize additional agency expenditures for priority programs and projects in excess of the original budget but only when revenue collections exceed the resource targets assumed in the budget or when additional foreign project loan proceeds are realized.

tools nationally (and for local government units and low income households) (section 3.2). There is evidence in the General Appropriation and the NDRRMF of funds appropriated and released for insurance premiums, but appropriation was reallocated in 2021, thus not actually spent. The 2018 ₱2.0 billion allocation was released to the Bureau of Treasury for parametric insurance premiums.

36.   **The National and District Disaster Risk Reduction and Management Fund utilization and responsiveness has continuously improved.** Several circulars were issued after Typhoon Yolanda in 2013[10] and a specialized manual on the receipt and utilization of funds during COVID-19 pandemic has been developed by the Commission of Audit. In general terms, to achieve full and timely use of the funds when needed, the commission has relaxed reallocation, procurement, and fund utilization rules during a calamity. The commission has also promulgated rules to be used in a declared emergency that are formalized in commission circulars to enable faster disbursements. This approach showed disbursement success, in that during the COVID-19 pandemic, the Commission of Audit reported to the consulting team that many funds were fully utilized.[11]

37.   **The fiscal losses resulting from disasters are cited in the Philippines' Fiscal Risks Statement.**[12] The statement is a multi-agency initiative, and the Fiscal Risks Statement Secretariat draws from the expertise of contributing agencies (including Department of Budget and Management and Department of Finance) to assess the economic and fiscal impact of disasters. Quantifying the risks accruing from disasters is mostly based on the historical average of damage and loss cited from reports issued by various agencies.

38.   **Participatory budgeting initiatives address disaster aspects.** During budget preparation, line agencies are directed to consult with civil society organizations and other stakeholders on ongoing and/or new spending/expansion agency programs and projects, accountability structure, costing, performance management, monitoring and evaluation, and risk management. The inputs from civil society organizations will then be reported and submitted with other budget preparation forms.

39.   **The existence of dedicated funds and explicit disaster-relevant budget appropriations for financing the impact of disasters highlights a high level of priority in this policy sphere.** The primary fund is the NDRRMF, also known as the Calamity Fund, as noted.

- The Calamity Fund was established in 2010[13] as a special purpose fund to support disaster prevention and mitigation, preparedness, response and recovery and rehabilitation activities.[14]

---

[10] Internationally known as Typhoon Haiyan.

[11] For COVID-19 related expenses the fund was, anecdotally, more easily disbursed. This is perhaps a function of the nature of cost behavior in that recurrent expense items are more readily acquired than infrastructure projects that require a design and lead time to come to the point of expenditure.

[12] Annual Fiscal Risk Statements from 2013 to 2022 can be accessed in the Fiscal Risk Statement (dbm.gov.ph)– Department of Budget and Management website (accessed 23 June 2022).

[13] Republic Act No. 10121. https://www.officialgazette.gov.ph/2010/05/27/republic-act-no-10121/.

[14] Department of Budget and Management. Calamity and Quick Response Funds. https://www.dbm.gov.ph/index. php/programs-projects/calamity-and-quick-response-funds#1-what-is-calamity-fund; and Status of National Disaster Risk Reduction and Management Fund https://www.dbm.gov.ph/index.php/programs-projects/status-of-national-disaster-risk-reduction-and-management-fund#2021.

- The fund's annual status reports (2018 up to 2021) indicate programmatic activity for:
  » the National Disaster Risk Reduction and Management Program;
  » quick response funds/augmentations to quick response funds[15] for national government and government-owned and controlled corporations;
  » Marawi Recovery, Rehabilitation and Reconstruction Program;
  » Comprehensive Aid to Repair Earthquake Damage; and
  » insurance premiums.[16]
- Over the 4 years reviewed, 29.9% of the value of releases from NDRRMF was to quick response funds (Table 5). It is understood that the intention of the NDRRMF was to provide 30% of its funds for relief and it can be seen that this has been achieved. Agencies holding a quick response funds need no further approval (other than declaration of a calamity) to spend funds released to them, so the intended agility and responsiveness has also been achieved.
- Since 2018, specific funded activities under the National Disaster Risk Reduction and Management Program accounted for 63.7% of appropriations and 68.0% of funds released to national government agencies and government-owned and controlled corporations.
- Within the National Disaster Risk Reduction and Management Program[17] (excluding quick response funds), Marawi Recovery, Rehabilitation and Reconstruction Program, and Comprehensive Aid to Repair Earthquake Damage funds were released for activity, including infrastructure for disaster recovery and for training of personnel, maintenance of assets and equipment, purchase of equipment, and repair of damage.
- Some releases for COVID-19 pandemic related programs were reported (for the Office of Civil Defense) in 2020 and 2021.
- Funds releases are approved by a Special Approval Release Order of the President of the Philippines who may take into consideration the recommendation of the National Disaster Risk Reduction and Management Council for local disasters or the appropriate agency for national crises.
- The NDRRMF twice had appropriations allocated for insurance premiums (in 2018 and 2021). The 2018 appropriation of ₱2.0 billion was released to the Bureau of the Treasury for Parametric Insurance Premiums (para. 83) but the 2021 appropriation of ₱2.0 billion was reallocated to the National Disaster Risk Reduction and Management Program in December 2021.
- The national quick response funds are built-in appropriations within the NDRRMF (accounted for within the National Disaster Risk Reduction and Management Program) that are pre-disaster or standby funds for national government agencies to enable rapid response to catastrophes and crises. The Department of Public Works and Highways, Department of National Defense—Office of the Secretary, Department of Health, Office of Civil Defense, Department of Education,

---

[15] Quick response funds are accounted for as part of the National Disaster Risk Reduction and Management Program.

[16] ₱2.0 billion allocated and released in 2018 and ₱2.0 billion allocated in 2021 but reallocated on 1 December 2021 to quick response funds replenishments and various projects. For information, see Annex D of the 2021 Accounts–and p. 1. See also https://www.dbm.gov.ph/index.php/programs-projects/status-of-national-disaster-risk-reduction-and-management-fund#2021 . 1/ Per OP approval dated December 1, 2021 the Insurance Premium is being realigned in view of critical balance of the National Disaster Risk Reduction and Management Program (p. 24 Annex F).

[17] For information, see Annex D of the Status of National Disaster Risk Reduction and Management Fund (dbm.gov.ph) National Disaster Risk Reduction and Management Program Report 31 December 2021.

Table 5: Programmatic Summary of National Disaster Risk Reduction and Management Fund Funding and Releases, 2018–2021

Type of Funding	Financial Year (₱ billion)				Totals	%
	2021	2020	2019	2018	...	...
Opening balance	5.135	6.226	5.142	0.000	...	...
National Disaster Risk Reduction and Management Program (including quick response funds and augmentations)	13.000	14.295	16.500	10.497	54.292	63.7%
Marawi Rehabilitation	5.000	3.500	3.500	10.000	22.000	25.8%
CARED	0.000	5.000	0.000	0.000	5.000	5.9%
Insurance Premiums	2.000	0.000	0.000	2.000	4.000	4.7%
Available Resources	25.135	29.021	25.142	22.497	85.292	100.0%
Released						
Insurance Premium	0.000	0.000	0.000	2.000	2.000	2.4%
NDRRMP (excluding quick response funds)	4.920	7.199	10.942	8.148	31.209	38.1%
Quick response funds (including augmentations)	10.155	9.234	2.789	2.325	24.503	29.9%
Marawi	4.670	6.929	4.463	4.882	20.944	25.6%
CARED	2.890	0.312	0.000	0.000	3.202	3.9%
Released	22.635	23.674	18.195	17.355	81.859	100.0%
Balance	2.500	5.347	6.947	5.142		
Earmarked	0.307	4.619	14.332	2.643		
Closing Balance	2.193	0.728	-7.385	2.499		

... = not available, CARED = Comprehensive Aid to Repair Earthquake Damage, NDRRMP = National Disaster Risk Reduction and Management Program.

Source: *Status of National Disaster Risk Reduction and Management Fund*. dbm.gov.ph (accessed 24 June 2022).

Department of Social Welfare and Development and Department of Agriculture have these built-in appropriations.

- A total quick response funds appropriation of ₱6.3 billion in 2021 and ₱6.4 billion in 2022 (part of the overall ₱20.0 billion) was proposed in the People's Budget to serve as a standby fund (or unprogrammed appropriations) in the following agencies in 2021: Department of Education (₱2.0 billion), Department of Social Welfare and Development (₱1.3 billion), Department of Agriculture (₱1.0 billion), Department of Public Works and Highways (₱1.0 billion), Department of Health (₱0.5 billion), and Office of Civil Defense (₱0.5 billion).
- In addition to these quick response funds, the NDRRMF financial statements indicate that quick response funds are held by government-owned and controlled corporations, including the National Electrification Administration, Philippines Port Authority, and Social Housing Finance Corporation.

- Local government units can apply for funding from the NDRRMF, and funds were released by the Department of Finance in 2020 and 2021 as part of the National Disaster Risk Reduction and Management Program to local government units. This is reported as assistance to the local government unit and related to Typhoons Quinta (Molave), Rolly (Goni), and Ulysees (Vamco).[18]

40.   **The NDRRMF is allocated and appropriated over and above regular budgets allocated to central and local government line agencies with mandates, projects, and activities contributing to climate actions and disaster resilience.** These line agencies were identified in the 2022 Fiscal Risks Statement (Department of Budget and Management 2020)[19] as the Department of Public Works and Highways, Department of Agriculture, National Irrigation Authority, Department of Transportation, Department of Energy, and Department of Labor and Employment.

41.   **Several direct or indirect disaster-relevant unprogrammed appropriations were set out in the 2021 and 2022 People's Budgets as standby appropriations.** These appropriations will become available once additional funding is acquired from surplus revenue collections, grants, or foreign loans, including ₱10.0 billion for COVID-19 Social Protection Programs within the "other" line (Department of Budget and Management 2021).[20] The use of programmed and unprogrammed appropriations for national government allocated directly to institutions and programs, as well as to special purpose funds on a regular and a reactive basis, is evidence of policy sensitivity to DRM and consequent responsiveness within budget allocation systems. Notably, budget reallocations within controlled rules are permitted to fund disaster response; these reallocations are permitted from underspends and are understood to be reversed at the end of the year and are returned to their original budgets, in keeping with the temporary nature of reallocations. Reallocations are not a formal part of the government's disaster risk financing, but this tool was used as part of COVID-19 response. The 2021 budget circular stated:[21] *"For 2020, the government will maintain disbursements at ₱4,175.2 billion (21.7% of GDP) to provide the additional health care, wage subsidies and other social safety nets intended to address the COVID-19 emergency through expenditure reallocation under the Bayanihan to Heal as One Act (R.A. No. 11469)."*

42.   **Historically, the Government of the Philippines has used supplementary budgets to fund disaster response.** In 2013, the Philippines was hit by a 1-in-30-year typhoon (Yolanda) which caused ₱93.4 billion in total losses and reduced the country's GDP growth by 0.6 percentage points. To fund the consequent relief, rehabilitation, and reconstruction, a supplemental budget was enacted to augment the NDRRMF by ₱11.2 billion and the Department of Social Welfare and Development's quick response funds by ₱3.4 billion. The government to some extent has also drawn on support from development partners to support both post-disaster recovery and reconstruction and COVID-19 response, including

---

[18] See Annex D, p. 16 of the December 2021 Fund Status Report. Under Special Provision No. 5 of the Local Government Support Fund—Disaster Rehabilitation and Reconstruction Assistance Program in fiscal year 2021 General Appropriations Act, Republic Act No. 11518, the appropriated amount of ₱4,500,000,000.00 shall be used to aid in the recovery and reconstruction of the economy and livelihood in the local government units directly affected by the Taal Volcano eruption, Typhoon Quinta, Super Typhoon Roily, and Typhoon Ulysses, as indicated in the situational reports issued by the National Disaster Risk Reduction and Management Council.

[19] See Fiscal-Risks-Statement-2022-for-Circulation.pdf. p. 75, dbm.gov.ph.

[20] See 2021 People's Budget, p. 10, final bullet. https://www.dbm.gov.ph/images/pdffiles/2021-Peoples-Proposed-Budget.pdf.

[21] See 2021 People's Budget, p. 4.

through a series of *ex-ante* contingent disaster financing loans from ADB, Japan International Cooperation Agency, and the World Bank. These include, for example, a \$500 million contingent disaster loan approved by ADB in 2019 disbursed in full in response to a series of typhoons hitting the country in October and November 2020 and the re-introduction of enhanced COVID-19 community quarantine arrangements in parts of the Philippines in March 2021.

43.   **Local government, like national government, has a mandate to provide DRM services to the population.** However, unlike national government, local government is mandated by Republic Act 10121 (Philippine Disaster Risk Reduction and Management Act of 2010) to set aside a prescribed level of resources. Not less than 5%[22] of the estimated revenue of local government units from their General Fund regular sources (DOF 2007),[23] including internal and external revenues and receipts comprising local taxation and non-tax revenues and the share of the National Tax Allotment, must be set aside as the Local Disaster Risk Reduction and Management Fund to be utilized in accordance with relevant DRM laws, rules, and regulations. If the *barangay* level needs more disaster financing, the municipal, city, and provincial governments allocate from their Local Disaster Risk Reduction and Management Funds.[24] This 5% allocation from the local government units' General Fund resources is further divided into 70% for preparedness and mitigation and 30% for quick response funds for immediate relief in similar fashion to the national arrangements. Local government units can accumulate unused funds in a special trust fund for up to 5 years providing ready funds for relief and for post-disaster purposes and also for disaster risk reduction. This rule is applied universally under the act and does not take local disaster risk into account. Pre-COVID-19 pandemic, a number of local government units had built up considerable resources in these special trust funds for future potential post-disaster purposes. However, due to the COVID-19 pandemic response, these funds have been depleted suggesting scope for extension of the 5-year period for accumulation to improve local resilience.

44.   **The National Tax Allotment is distributed to local government units based on a reasonably simple three-stage formula.** Stage 1 allocates a percentage of the National Tax Allotment to each tier of government as set out in Table 6. There are 4 levels of local government units: barangay (lowest), municipal, city, and provincial governments. Stage 2 allocates the National Tax Allotment to individual provinces, cities and municipalities based on the formula set out in Table 7. The third and final stage is to calculate the share for each barangay which is set at ₱80,000 for each unit with a population of at least one hundred and the balance allocated 60% based on population and 40% on an equal sharing basis.

45.   **The redistribution of resources from national to local government has the potential to fragment resources available for and allocated to DRM.** As of 2022, the share of national funds allocated to local government units will increase significantly due to the redistributive effects of the Mandanas Ruling combined with the mandated requirement to allocate not less than 5% of the estimated revenue from regular sources to the Local Disaster Risk Reduction and Management Fund. The ruling should result in a higher amount

---

[22] National Disaster Risk Reduction and Management Council-DBM-DILG .JMC No. 2013–1 dated 25 March 2013 as referenced in the DBM Local Budget Memorandum (No 82 Dated 14 June 2021)–See Paragraph 2.2.5.3 on p. 6; also see Section 324d of RA7160. https://www.officialgazette.gov.ph/1991/10/10/republic-act-no-7160/.

[23] In DOF (2007), (for definitions of "regular sources" see p. 12—confirms that regular sources incudes the internal revenue allotment—now renamed the national tax allotment) (Microsoft Word - SIE CY 2004 Volume 1z.doc [blgf.gov.ph]).

[24] The *barangay* is the lowest level of government, referring to a village, district, or ward.

Table 6: Allocations of National Tax Allotment to Tiers of Local Government

Local Government Unit	Allocation of National Tax Allotment
Provinces	23%
Cities	23%
Municipalities	34%
Barangays	20%
Total	**100%**

Source: Republic Act No. 7160. Section 285. https://www.officialgazette.gov.ph/1991/10/10/republic-act-no-7160/.

Table 7: Basis of Distribution of National Tax Allotment to Individual Local Government Units

Factor	Percentage	Source of Data
Population	50%	Census of Population as declared official for all purposes through a Presidential Proclamation[a]
Land Area	25%	Official and Validated Master List of Land Area[b]
Equal Sharing	25%	
Total	**100%**	

[a] As per the Philippine Statistics Authority.
[b] As per the Lands Management Bureau.
Source: Authors.

dedicated to DRM at the local government unit level but could result in lower sums available for allocation to DRM by national government. Following the Mandanas ruling[25] by the Philippine Supreme Court in 2018 (and confirmed in 2019) the National Tax Allotment allocated to local government units is proposed to increase to ₱959.0 billion in 2022[26] which is ₱263.5 billion, or 37.9%, higher than in 2021. Essentially, the effect of the Mandanas ruling is to allocate to local governments a share of all national internal revenues—not only those collected by the Bureau of Internal Revenue but also collections by the Bureau of Customs. With this change, local government units will receive a higher share of national taxes. In the 2022 national budget this allocation is around 4.8% of GDP in 2022 compared to 3.5% in 2021 (World Bank 2021). The local government unit share of national resources is classified as an automatic appropriation (Department of Budget and Management 2020)[27] with consequential impact on the quantum available to local government units for DRM. Integration of, and planning for the effective use of, an increase in resources of this magnitude is a potential change management issue that may require direction and leadership from the Department of Budget and Management, the Commission of Audit and the Department of Interior and Local Government.

---

[25] See the Center for Local and Constitutional Reform (2020) for information on the ruling.
[26] 2022 People's Budget, p. 5.
[27] See p. 20 of the People's Budget (2022) references to Automatic Appropriations and as per Republic Act No. 9358, appropriating a supplemental budget for 2006 states that the Internal Revenue Allotment (now the National Tax Allotment) is considered *automatically appropriated* and that future local government share in the national taxes shall be *automatically appropriated* (section 4).

46.   In dialogue with the consulting team,[28] the National Economic Development Agency reported that the current computation of the National Tax Allotment is not risk or needs sensitive and recognized that change to the code is accordingly needed. This is a significant task that will require considerable consultation and time to complete—nevertheless, the scale of the challenge is not a sound reason to avoid it. This initiative would address a gap, enhancing an already established allocative calculation. Non-ring-fenced resource allocations to local government units influenced by rational disaster risk do not guarantee sums will be allocated to this policy domain so consideration should be given to establishing or utilizing special or matching grants to incentivize resource allocation to DRM.

47.   **Several other sources of DRF are available to local government units largely funded by national government (Box 3).** These include allocations from the NDRRMF based on requests for support and subsequent post-disaster needs assessments conducted by government. There were some 350 requests in 2021 for additional disaster funds from local government units. At the time of the fieldwork stage of this study in November 2021, the Office of Civil Defense had approved more than 20 applications. Local government units also reported using their general fund resources (as appropriated directly and using temporary reallocations) for recovery. This supplements Local Disaster Risk Reduction and Management Fund allocations and suggests scope for a review of trust rules that place a time limit on the accumulation of funds from the Local Disaster Risk Reduction and Management Fund and consideration of a shift to rules that are more risk-led than universally applied. This review could focus on establishing a disaster risk led approach with a more flexible approach on carry forwards and retained balances considered for higher risk local government units. If successful, this would help to address the risk of reduced allocations to local funds that arise from reduced local income sources post disaster and also the frequently-reported risk that local quick response funds are quickly exhausted by disaster events.

48.   **Local government units are also mandated to appropriate 20% of their National Tax Allotment distribution to their Local Development Fund for development projects.**[29] Copies of the local government unit development plans are provided to the Department of Interior and Local Government[30] and it was reported to the consulting team that this fund had been used for post-disaster recovery and resilience in infrastructure.

49.   **The government is evaluating the use of risk analysis to locate the most vulnerable local government units.** In discussion with the Office of Civil Defense[31] it was reported to the consulting team that the secretariat is trying to use risk analysis to locate the most vulnerable local government units. The consultant team sees the availability of data relating to disaster risks as an opportunity to match National Tax Allotment distributions more closely with disaster-related needs. Given the mandatory nature of the 5% allocation for DRM, any reflection of disaster risk reduction within the allotment will result in closer matching of DRM resources with risk.

---

[28] At the workshop in Manila titled Workshop on Philippines Disaster Risk Financing Diagnostic Assessment on 5 October 2022.

[29] Department of Interior and Local Government and Department of Budget and Management, Joint Memorandum Circular 2017–1, 22 February 2017. References Republic Act 7160 Paragraph 287.

[30] RA 7160. https://www.officialgazette.gov.ph/1991/10/10/republic-act-no-7160/.

[31] The secretariat of the National Disaster Risk Reduction Council.

Box 3: Local Government Sources for Disaster Financing

- Local Government Support Fund[a]
- Assistance to Municipalities from the National Disaster Risk Reduction and Management Fund
- Assistance to Cities from the National Disaster Risk Reduction and Management Fund
- Conditional Matching Grant to Provinces for Road and Bridge Rehabilitation, Upgrading and Improvement
- Provision for Potable Water Supply (SALINTUBIG)
- Other Financial Assistance to Local Government Units from National Government
- Disaster Rehabilitation and Reconstruction Assistance Program (a constitutional right of the local government units)

[a] Allocation of this fund has already been broken down into specific agencies and/or programs and projects but are presented as funds in the budget. See 2020 People's Budget, p. 11. https://www.dbm.gov.ph/images/pdffiles/2020-Peoples-Proposed-Budget.pdf.

Source: Authors.

2.3 Diagnostic and Recommended Actions

50.   **Appropriation for the NDRRMF is determined annually, compromising long term planning.** NDRRMF appropriation stood consistently at approximately ₱20 billion in fiscal year (FY) 2018 to FY2022 despite significant fiscal pressures. However, this was not the case in the past, with allocations to the NDRRMF varying from ₱3.75 billion in FY2010 to a peak of ₱38.9 billion in FY2016. The budget increased again in FY2023 to ₱30 billion as part of government efforts to take a more proactive approach in building disaster resilience.

51.   In contrast, the National Tax Allotment is an automatic appropriation for local government units as confirmed by the Mandanas Ruling.[32] Local government units, in turn, are mandated to allocate funding for DRM on a given minimum percentage (5%) thus assuring predictable and protected funding for disaster risk reduction and management.[33]

> **Funding for the national government NDRRMF should be considered for automatic appropriation status based on a percentage of GDP or other formula to be determined.** *If this policy, with legal force, was applied to the NDRRMF the fund would be guaranteed a predictable level of income every year in the same way as local disaster risk reduction and management funds.*

52.   **The local special purpose disaster funds have not been fully used in the past.** The special purpose disaster funds are a welcome instrument for both national and local government and are well used by national government. But local governments did not always use them until the COVID-19 pandemic. This suggests the need for a reorganization of DRF

---

[32] See p. 20 of the People's Budget (2022), references to automatic appropriations. And as per Republic Act No. 9358, appropriating a supplemental budget for 2006 states that the Internal Revenue Allotment (now the National Tax Allotment) is considered *automatically appropriated* and that future local government share in national taxes shall be *automatically appropriated* (section 4).

[33] National Disaster Risk Reduction and Management Council-DBM-DILG .JMC No. 2013–1 dated 25 March 2013 as referenced in the DBM Local Budget Memorandum (No 82 Dated 14 June 2021). See paragraph 2.2.5.3, p. 6; see also RA 7160, section 324d. https://www.officialgazette.gov.ph/1991/10/10/republic-act-no-7160/.

from public sector sources. At the same time, the full and timely use of the funds to cope with a disaster, both *ex ante* and *ex post* should be enhanced. Another factor that may explain why local governments underused funds pre-COVID-19 is that local government units had often deliberately under spent on the 70% share for prevention with the intention to build up resources for use in the event of a potential disaster (paragraph 43)—i.e., for response. As such, the Local Disaster Risk Reduction and Management Fund design as a single pot contributes to the "underspend" on disaster risk reduction and so fails to help reduce underlying risk or, ultimately, disaster damage and loss.

53.   **Consideration should be given to guiding local government units on the use of the increased resources that will result from the Mandanas Ruling.** *The redistributive effect of the ruling will result in more resources automatically allocated for DRM by each local government unit.* The Bureau of the Treasury indicated to the consultant team that following implementation of the Mandanas ruling, the local government would be expected to take on greater responsibility for disaster risk financing. The following should be done:

- Local government units should be incentivized to spend more of this new resource on proactive Disaster Risk Reduction and Risk Transfer solutions and also to consider arrangements for pooling quick response funds.
- There is also scope for review of Local Disaster Risk Reduction and Management Fund trust and carry forward rules to be more risk-led rather than universally applied.
- In addition, a separation of the Local Disaster Risk Reduction and Management Fund for disaster risk reduction and preparedness and meeting of post-disaster spending demands should be considered.

54.   **The National Tax Allotment distribution formula for local government (and consequently the sum allocated for DRM) does not recognize the impact of disaster risks other than at a general level of population and land area.** For example, in some local government units significant transient populations affect relief or risk reduction needs, or disaster risk is greater (e.g., in coastal areas) which may require more investment in risk reduction and preparedness and greater resources for reconstruction. The consulting team was also told that some unease exists among local government units that recent disaster losses were not reflected in the distribution of the National Tax Allotment or by way of special grants on a one-off basis.

55.   **Levels of disaster risk and recent losses are potentially significant given that each disaster can reduce local government unit revenues.** This has an immediate impact on the required 5% contribution to the Local Disaster Risk Reduction and Management Fund,[34] while placing additional expenditure demands on separately mandated disaster-related activities. Typically, this is *"to be in the frontline of service delivery in the immediate relief during and assistance in the aftermath of man-made and natural disaster and natural calamities."* In addition, it is more proactive activity as set out in law (Republic Act 7160 The Local Government Code). This includes activities assigned to several governing bodies. These include the Office of the Mayor, the Sangguniang Panlungsod (the legislative body),

---

[34] The Commission of Audit indicated that the 5% local disaster risk reduction and management fund is based on the local government units estimated revenue/allocation for the following year. In some cases, the revenue is not reached, resulting in lower than anticipated credits to the fund. Local government units thus carry the risk that disasters or other events impact on the sums available to the local fund. The redistributive effect of Mandanas will mitigate this, but the risk remains.

the Administrative Office, the Legal Office, the Agriculture Office, the Social Welfare and Development Office, the Environment and Natural Resources Office, the Planning Office, the Information Office, the Cooperatives Office, the Veterinary Office and the General Service Office.

56.   **Local government units are classified according to their average annual income from regular source[35] in the most recent 4 years (1 = largest; 6 = smallest).[36]** This classification has no *direct* impact on the National Tax Allotment distribution formula. However, it does influence the maximum taxable ceilings, the determination of special and specific grants, salary scales, and other aspects of finance and human resources, including ability to undertake development projects.[37] It was reported to the consulting team that the current distribution formula is perceived as adversely affecting disaster-prone smaller authorities in classes 4 to 6 that have low local revenues (which are at risk from disasters) and proportionately higher fixed costs that are unable to take advantage of economies of scale.

57.   **There is differentiated climate risk and disaster risk across the country, which further suggests merit in recognizing this in the National Tax Allotment distribution formula and in the operational rules of local special purpose disaster funds.** For example, the National Integrated Climate Change Database and Information Exchange System[38] identifies coastal areas as vulnerable to particular risks. The National Development Plan, 2017–2022, recognizes the importance of "local" in this regard:

> "To increase adaptive capacities and resilience of ecosystems *strengthen the implementation of climate change adaptation and disaster risk reduction across sectors particularly at the local level strategies are best formulated at the regional and local levels."*[39]

58.   **It is therefore recommended that the National Tax Allotment distribution formula is reviewed.** The revision should consider recognizing the differentiated risk of disasters and thereby support local government units financially to take their own allocative decisions to manage disaster risks locally:

> **The government should consider reviewing the National Tax Allotment methodology to take account of relative climate change and disaster risk.** *The distribution of the National Tax Allotment between local government units is driven in large measure (75%) by land area and population with the balance divided equally within each tier of local government. The distribution formula does not differentiate between levels of disaster or climate change risks, or recognize recent disaster losses. This approach would match resources to disaster risk and related needs more closely.*

---

[35] Share of *all* taxes collected by the Bureau of Internal Revenue, *Bureau of Customs*, and other agencies as per the Mandanas Ruling plus local revenues.

[36] Executive Order No. 247, s.1987 Providing for a New Income Classification of Provinces, Cities and Municipalities, and for other Purposes, section 1 https://www.officialgazette.gov.ph/1987/07/25/executive-order-no-249-s-1987/.

[37] Executive Order No. 247, s.1987, section 5 https://www.officialgazette.gov.ph/1987/07/25/executive-order-no-249-s-1987/.

[38] National Integrated Climate Change Database and Information Exchange System, https://niccdies.climate.gov.ph/adaptation/coastal (accessed June 2021).

[39] Philippines Development Plan (National Economic Development Agency) 2017–2022, p. 42.

59.   **The government's annual Fiscal Risks Statements identify disasters as a major source of fiscal risk, but a centralized register of disaster-related contingent liabilities is not fully operational.**[40] The statement sets out the costs of losses and damage reported to government for each event, without explicitly identifying the total costs to government considering the local and national disaster risk reduction and management funds and other expenditure.[41] It was also reported to the consulting team that there is no centralized register of disaster-related contingent liabilities. An initiative to track expenditures on an event-by-event basis could be a key enabler of centralized registration of disaster related contingent liabilities and fuller compliance with the government's adopted International Public Sector Accounting Standards (IPSAS) 19 (Provisions, Contingent Liabilities and Contingent Assets) standard (International Federation of Accountants 2019).[42] The 2019 Open Budget Survey, in which the Philippines scored very highly for transparency,[43] identifies that while contingent liabilities relating to government-owned corporations and debt are reported, fiscal risks relating to disasters are not.[44] Although encouraging, the government has a classification within contingent liabilities for this purpose.

60.   **The Unified Chart of Accounts**[45] **is used by all government institutions to classify budgets and expenditure.** It has functionality to classify and aggregate budgets or expenditure across programs/projects and line items (known as object codes) across the whole of government. This functionality could be readily adapted, using a temporary code or label, to tag and aggregate government expenditure relating to individual events or declared calamities and in turn enable more comprehensive preparation and reporting of government-wide disaster-related fiscal risks. A total of 25 disaster-related policy codes were identified within Unified Chart of Accounts, but none operate specifically at event level. Instead, as designed and in line with classification norms, these identify policy drivers of expenditure. This is a challenging area that will require time and attention to evolve to fruition. The development of financial data and valuation systems, as well as the application of evidence-based judgements, will be key in the longer term. Examples of how to potentially begin this exercise could include continued separation of government and nongovernment damage and losses in post-disaster needs assessments and a rationalization of major final output codes to identify components of DRM, based on Sendai for example, as well as event-based, temporary "tagging" of expenditure within mainstream and regular appropriations on a multi-agency basis. These approaches, linked and integrated with data that exists in the Philippines Statistics Authority and Office of Civil Defense for example, would generate classified financial data that could be refined over time to begin to value implicit contingent liabilities faced by the government.

---

[40] 2020 Fiscal Risk Statement, p. 2. FY-2020-Fiscal-Risks-Statement.pdf (dbm.gov.ph).

[41] 2020 Fiscal Risk Statement, See p. 59 for examples.FY-2020-Fiscal-Risks-Statement.pdf (dbm.gov.ph).

[42] See Case-Study-Adoption-of-IPSAS-Philippines.pdf (ifac.org).

[43] International Budget Partnership. Open Budget Survey: Rankings. 2019. See Rankings | International Budget Partnership.

[44] The classifications are (i) national government direct guarantees (debt guarantees to government owned or controlled corporations; indirect guarantees (public–private partnerships); (ii) contractual obligations from public–private partnerships; and (iii) risks arising from natural disasters [*** see recommendation on term above] and calamities (implicit contingent liabilities, i.e., those that represent moral obligations or burdens that, although not legally binding, are likely to be borne by governments. because of public expectations or political pressures [IMF 1999]).

[45] See Unified Chart of Accounts at https://www.uacs.gov.ph/resources/faqs for further information.

The government should establish a central register of contingent liabilities that identifies and values disaster-related fiscal risks. *While a disaster risk sub-classification is already established within the government's contingent liability classification, a centralized fiscal risk register incorporating disasters is needed. There are multiple rich sources of data available from the Philippines Statistics Authority and Office of Civil Defense to help with this.*

61.　　**Financing for DRM exists beyond the main program and the NDRRMF in potentially significant amounts in mainstream, regular budgets.** The Department of Public Works and Highways and the Department of Social Welfare and Development utilize their mainstreamed, regular appropriations for DRM activities. The Department of Public Works and Highways uses maintenance budgets on a risk-led basis but budgets are not allocated to the agency with any recognition of disaster risk. The Department of Social Welfare and Development reported utilizing a disaster "fund" financed from the agency's appropriation and "top ups" from social protection programs. A "Diaspora Fund" under the Department of Foreign Affairs also accepts donations as part of relief financing. These financing mechanisms should formally be quantified and recognized within the government's DRF strategy.

The government should consider conducting a Disaster Risk Management Public Expenditure and Institutional Review. *Technical and jurisdictional scope should be determined but transparency of the scope and quantum of DRF resources nationally should ultimately be known. This should aim to identify the roles and resources outside the National Disaster Risk Reduction and Management Fund Program in Disaster Risk Management and to inform a "refresh" of the Disaster Risk Financing Strategy.*

62.　　**The 2019 Open Budget Survey identified a gap in registration, valuation, and disclosure of disaster-related fiscal risks and contingent liabilities.**[46] The Unified Chart of Accounts has functionality to identify and aggregate whole of government expenditure of individual disaster events. The government should investigate the use of temporary expenditure tags under the project/program dimension of Unified Chart of Accounts to support identification of costs to government of individual events and in turn improve preparation for the fiscal impact of disasters.

The government should investigate the use of temporary expenditure markers to identify the costs of disasters at event level. *This could be done under the project/program dimension of Unified Chart of Accounts to support identification of costs to government of individual events and in turn improve preparation for the fiscal impact of disasters. This financial information will support the establishment of registers of contingent liabilities.*

---

[46] International Budget Partnership. Open Budget Survey: Philippines 2019. https://internationalbudget.org/open-budget-survey/country-results/2019/philippines.

63.   **During the COVID-19 pandemic, clarity was insufficient about the role of the national government versus the role of local government units in responding to the pandemic, particularly in relation to planning and financing the health response.** Following the World Health Organization (WHO) declaration of a pandemic, and amid rising cases of COVID-19 in the Philippines, proclamation number 92 was issued declaring a national state of emergency.[47] Congress then convened a special session to pass the "Bayanihan to Heal as One Act."[48] The act provided additional powers to the President to (i) "reallocate, realign, and reprogram" a budget of around ₱275 billion ($5.37 billion) from the estimated ₱438 billion ($8.55 billion) national budget approved for 2020; (ii) "temporarily take over or direct the operations" of public utilities and privately owned health facilities and other necessary facilities "when the public interest so requires" for quarantine, the accommodation of health professionals, and the distribution and storage of medical relief; and (iii) "facilitate and streamline" the accreditation of testing kits. Combined with the nationally led response of the Inter-Agency Task Force for the Management of Emerging Infectious Diseases (IATF-EID), this resulted in tension with some local government units and confusion on roles, responsibilities and budgetary decisions—including decisions on declaring states of emergency and managing the direct health response. This is because under normal circumstances, the management of the health system is heavily decentralized with funding for local public hospitals and healthcare providers at the discretion of local government units.

The policy on health system decentralization needs additional details on roles and responsibilities, especially with respect to financing health-related disasters. *Government policy needs to include specific identification of the roles and responsibilities of national government versus local government units at times of major health-related disasters, particularly during communicable disease outbreaks.*

---

[47] President of the Philippines, *Proclamation 922 Declaring a state of public health emergency throughout the Philippines*, 2020 Proclamation No. 922 s. 2020 | Official Gazette of the Republic of the Philippines.
[48] Government of the Philippines, Republic Act 11469, Bayanihan-to-Heal-as-One-Act-RA-11469.pdf (senate.gov.ph).

Diagnostic on the Current Availability and Utilization of Insurance, Reinsurance, and Capital Markets for Disaster Risk Financing

3.1 Economic Conditions and Other Support Functions

3.1.1 Economic Landscape

64.   **After an unprecedent recession leading to a 9.6% decrease in real GDP during 2020, the Philippines is recovering from COVID-19.** GDP in 2021 grew 5.7% and the Asian Development Outlook (ADB 2022a) forecast growth for the country of 7.6% in 2022 (ADB 2023) and 6.3% in 2023 (ADB 2022a).

65.   **However, important challenges remain that could increase the financial vulnerability of the country to natural hazards, epidemics, and pandemics in the next few years, exacerbating the social and economic impacts of such shocks:**

- Almost 10 million are estimated to have fallen into poverty due to the pandemic. [49]
- Uncertainties regarding the remaining course of the pandemic and the emergence of new coronavirus variants across the globe. The Philippines' COVID-19 vaccine rollout may suffer from global supply shortages in the short term and local community quarantines could be extended to curb the spread of COVID-19.
- Inflation rose to 3.9% in 2021, up from 2.4% in 2020. *"Supply-side pressures from adverse weather and African swine fever pushed food prices higher. Domestic petroleum prices rose in line with global oil price trends. The government took temporary measures last year to augment food supplies, including reducing rice and meat import tariffs and allowing more pork imports"* (ADB 2022a). However, inflation is expected to ease to 3.5% only in 2023 as the government takes measures to address supply-side pressures (ADB 2022a).
- Merchandise exports are expected to increase with the rise in global trade, as imports, especially capital goods, rebound to support public infrastructure development (ADB 2021b).

---

[49] "Compared to the baseline path for poverty, the 2020 figure is 144 million people higher. Some of this will be offset as economies start to recover in 2021, but the longer-term scenario suggests that half of the rise in poverty could be permanent. By 2030, the poverty numbers could still be higher than the baseline by 60 million people." https://www.asianjournal.com/philippines/across-the-islands/ph-among-countries-to-see-an-increase-in-extreme-poverty-due-to-covid-19/ citation from Brookings Institution https://www.brookings.edu.

- The unprecedented return of overseas Filipino workers to the Philippines has posed growing challenges, particularly the implications for remittances, income, and employment.[50] As a major migrant-sending and remittance-receiving economy, the emigration of workers has been part of the Philippines' development narrative, and the COVID-19 impact reduction measures had severe impact on that position.

66.   **A shortage of physicians and nurses and insufficient supply of resources across the economy limits capacity to respond to pandemics and epidemics.** Just 0.6 doctors and 4.9 nurses serve every 1,000 people (Figure 4), less than a quarter and just over half, respectively, of the Organisation for Economic Co-operation and Development (OECD) average (Figure 5). Healthcare resources, including critical infrastructure and workforce, are also concentrated in the capital, with many parts of the economy still lacking access to quality facilities and personnel. The private sector, which can be accessed by a small percentage of the population, has an increasing number of high-quality facilities, but these are typically geared to elective surgical work and diagnostics (CT scans, MRI, etc.) and not well equipped to deal with communicable diseases and the effects of pandemics and epidemics.

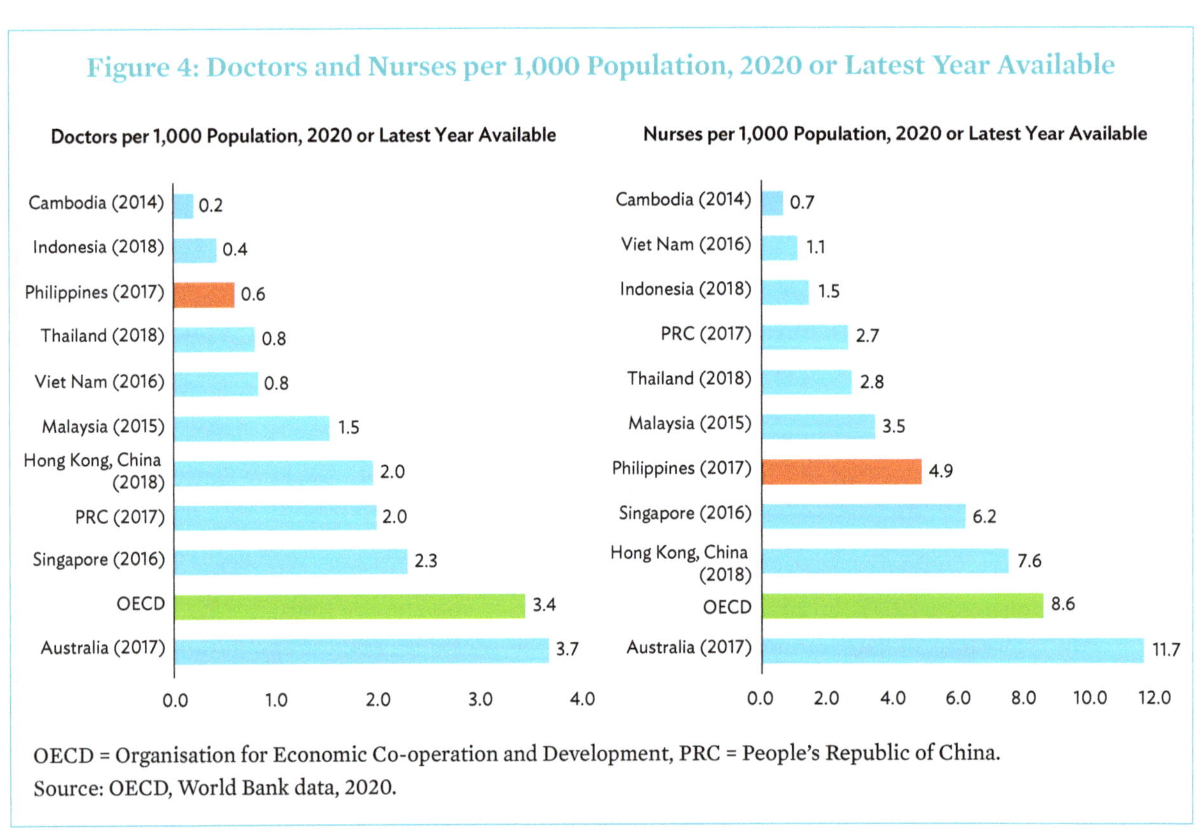

Figure 4: Doctors and Nurses per 1,000 Population, 2020 or Latest Year Available

OECD = Organisation for Economic Co-operation and Development, PRC = People's Republic of China.
Source: OECD, World Bank data, 2020.

[50] The ADB study, *COVID-19 and Overseas Filipino Workers Return Migration and Reintegration into the Home Country—The Philippine Case* suggests policy insights to improve and strengthen the pool of return and reintegration strategies to better leverage the country's outmigration propensity as a tool for economic transformation. See COVID-19 and Overseas Filipino Workers. https://www.adb.org/sites/default/files/publication/767846/sewp-021-covid-19-ofws-return-migration-reintegration.pdf. The economic recession from the COVID-19 pandemic threatened the job security and well-being of over 91 million international migrants from Asia and the Pacific. See COVID-19 Impact on International Migration, Remittances, and Recipient Households in Developing Asia. *ADB Briefs*. https://www.adb.org/sites/default/files/publication/622796/covid-19-impact-migration-remittances-asia.pdf.

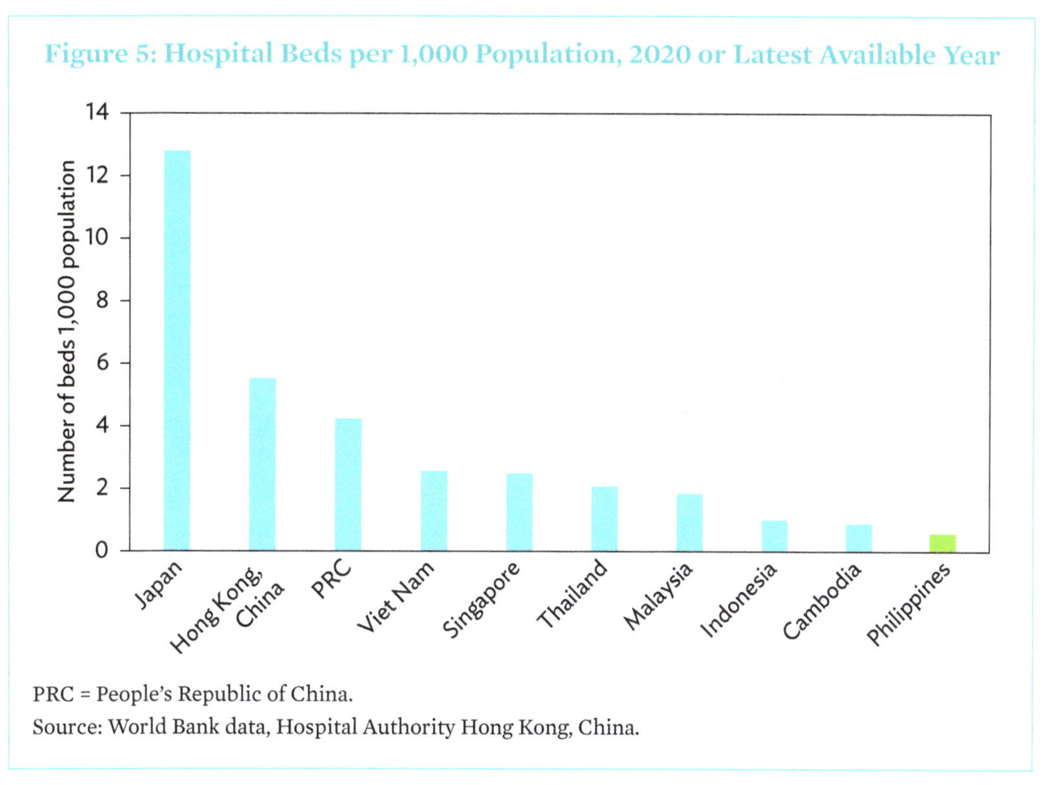

Figure 5: Hospital Beds per 1,000 Population, 2020 or Latest Available Year

PRC = People's Republic of China.
Source: World Bank data, Hospital Authority Hong Kong, China.

3.1.2 Data Availability

67.   **The Unified Chart of Accounts lists the classification codes for both infrastructure and hazard-exposed fixed assets, but comprehensive registration and valuation is still being developed.** The government fixed asset register is being implemented by the Bureau of the Treasury. Gaps in registration and valuation are reported in the Commission of Audit annual audit reports for both national and local government.[51] Asset valuation in the Bureau of the Treasury is consistent with historic cost under the applicable (IPSAS-17) accounting standard that also allows for valuation based on the fair value concept.[52] Insurance valuations are based on replacement cost which requires a regular appraisal to remain up to date—IPSAS 17 recommends 3–5 yearly revaluations.[53] It was reported to the consulting team that revaluations have not yet taken place.

---

[51] See Department of Education as an example; Northern Samar https://www.coa.gov.ph/phocadownloadpap/userupload/annual_audit_report/local   government   units/2020/Region-VIII/Provinces/NorthernSamarProv_ES2020.pdf https://www.coa.gov.ph/index.php/national-government-agencies/2020/category/8997-department -of-education. https://www.coa.gov.ph/phocadownloadpap/userupload/annual_audit_report/local government units/2020/Region-VIII/Provinces/NorthernSamarProv_ES2020.pdf sets out similar issues.

[52] International Public Sector Accounting Standard, IPSAS 17 - Property, Plant and Equipment https://www.ifac. org/system/files/publications/files/ipsas-17-property-plant-2.pdf.

[53] IPSAS 17- Property, Plant and Equipment, p. 527, para. 49.

68.   **Several agencies are engaged in collecting pertinent disaster data.** The Philippines Institute of Volcanology and Seismology (PHIVOLCS) and the Philippine Atmospheric, Geophysical and Astronomical Services Administration (PAGASA), both under the Department of Science and Technology are the apex institutions for monitoring, data collection and dissemination of meteorological, hydrological, and geophysical disasters in the country.

- PHIVOLCS has a network of sensors and field stations across the country to monitor seismological and volcanological activity. It also has doppler radars for accurate capturing of seismological data and soil moisture sensors. PHIVOLCS collects raw data but does not initiate data modeling independently. It has conducted customized hazard assessments for banks and has a memorandum of understanding with the Philippine Insurers and Reinsurers Association (PIRA) for development of risk models.
- PAGASA has a network of observatories, radars, ground stations, and automatic weather stations to measure wind velocity, wind direction, and rainfall across the country, providing flood warnings and extreme weather predictions. It has almost 50 years of data on wind and precipitation. However, issues exist in the functioning of some automatic weather stations and the consistency and accuracy of data they generated.
- Apart from PHIVOLCS and PAGASA, other agencies collect significant ground level data on disaster damage and loss, beneficiaries, productivity, climate change impact, etc., For example, the Department of Agriculture administers several agriculture credit and extension programs for farmers that collect a wide range of data. The Department of Agriculture also collects granular crop loss data following every disaster. Similarly, the Department of Agrarian Reform administers several schemes for agrarian reform beneficiaries and has data on their economic conditions and agriculture practices (section 3.2.5).

69.   **Disaster risk modeling for the Philippines, while still incomplete, has advanced.** The Oasis Loss Modeling Framework, the Reinsurance Corporation of Philippines, and PAGASA, among other partners, have collaborated on the development of an open-source flood risk model for the Philippines. Modeling firms AIR Worldwide (now VERISK) and Risk Management Models, Analytics, Software and Services both have proprietary earthquake and typhoon models for the Philippines.

70.   **The Philippine Crop Insurance Corporation (PCIC) has gathered a large amount of historical data on agriculture risks.** Established in 1978, PCIC has been offering agriculture insurance in the Philippines for several decades. It has gathered a large amount of data that can be analyzed for establishing loss patterns, cropping patterns, land use and ownership, production records, profit margins, and so on. Other agencies such as the Department of Agriculture, the Department of Agrarian Reform, Department of Finance, Land Bank of the Philippines, the Development Bank of the Philippines, the Department of Environment and Natural Resources, and the National Disaster Risk Reduction and Management Council also collect and maintain huge amounts of data on agriculture. However, the collected data would need significant cleaning, mining, and manipulating before actuaries, insurers, banks, agriculture service providers, etc., could effectively use it for services and enable development of specific risk models for agricultural insurance.

71.   **The Philippine Integrated Disease Surveillance and Response (PIDSR) System meets International Health Regulations on pandemics and epidemics**. The International Health Regulations are the legal framework *"to prevent, protect against, control and provide a public health response to the international spread of disease in ways that are commensurate with and restricted to public health risks, and which avoid unnecessary interference with international traffic and trade"* (WHO 2016). The International Health Regulations Monitoring and Evaluation Framework is used to *"review the implementation of country core public health capacities under the International Health Regulations (2015) and provides opportunities for continuous improvement"*(WHO 2018). The last Joint External Evaluation was conducted in 2018 (WHO 2018), and noted that considerable progress had been made in bolstering the country's public health surveillance, particularly at the national level. The evaluation highlighted the development of a Field Epidemiology Training Program which "strengthened health security in the Philippines by training a cadre of professionals for the core public health functions of surveillance, risk assessment and response at the local, regional and national levels" (WHO 2018). The development of the PIDSR and efforts to integrate this system across all levels of care and government and across all relevant communicable disease types were also commended. The conceptual framework for the PIDSR (Figure 6) highlights the system's position within the overall health and broader emergency response systems.

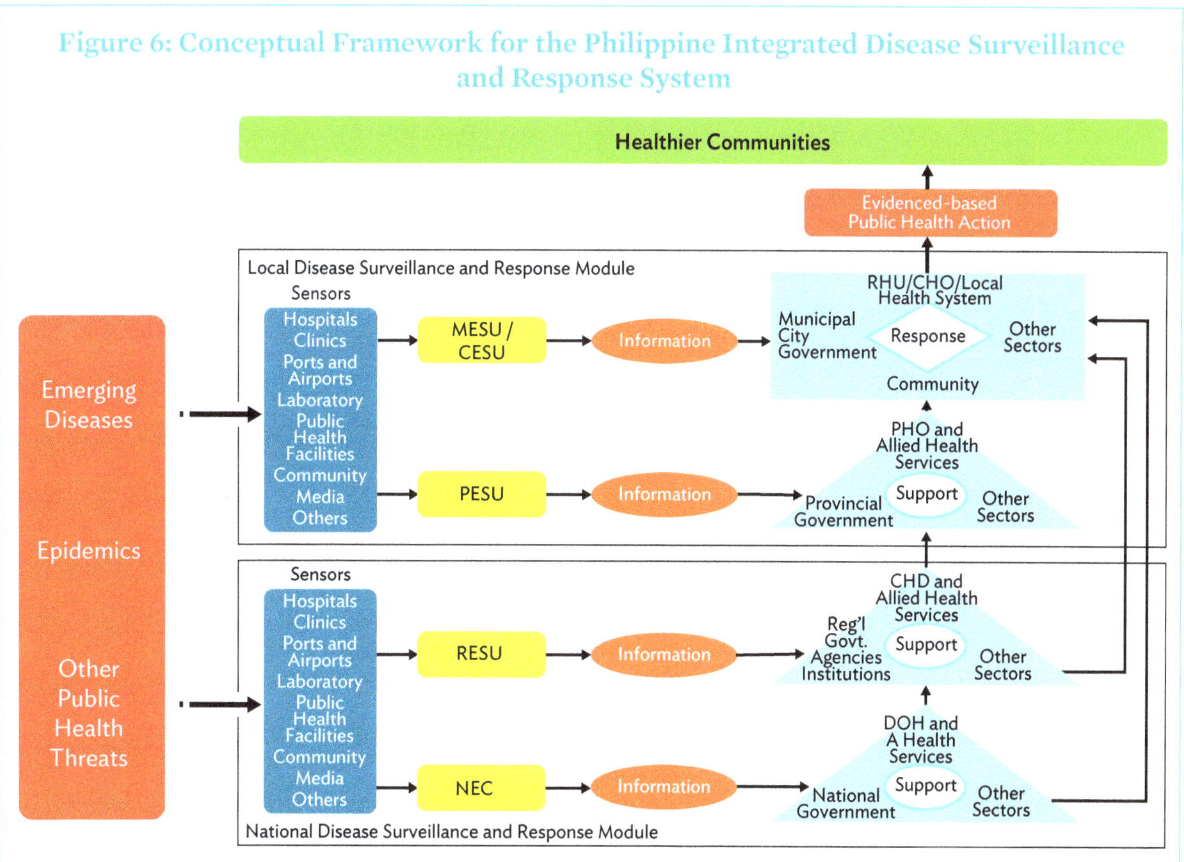

Figure 6: Conceptual Framework for the Philippine Integrated Disease Surveillance and Response System

CESU = City Epidemiology and Surveillance Unit, CHD = Center for Health Development, CHO = City Health Office, DOH = Department of Health, MESU = Municipal Epidemiology and Surveillance Unit, NEC = National Epidemiology Center, PESU = Provincial Epidemiology and Surveillance Unit, PHO = Provincial Health Office, RESU = Regional Epidemiology and Surveillance Unit, RHU = Rural Health Unit.

Source: Government of the Philippines, Department of Health. 2014.

72.   **While the PIDSR is in place, accuracy, completeness, and timeliness of the data captured is an ongoing challenge, which limits effectiveness in real-time epidemic and pandemic risk surveillance.** A number of other surveillance systems are not integrated into the PIDSR. The Philippines has both indicator-based surveillance systems (e.g., notifiable diseases) and event-based surveillance systems (e.g., an epidemic or pandemic), but these are not fully integrated and rationalized. There are several indicator-based surveillance systems, with PIDSR being the main system used. The PIDSR system includes 19 notifiable diseases, principally related to known diseases such as tuberculosis. There is also a syndromic surveillance system for other outbreaks, including encephalitis syndrome, severe acute respiratory infection, and influenza-like-illness. All systems use electronic data generated by public health facilities. Private health facilities are currently excluded from the system, which is a limitation, particularly as private hospital beds account for a significant percentage of total hospital beds. Data is captured by local Epidemiology and Surveillance Units and sent to the Regional Epidemiology and Surveillance Units for collation, before sending to the Department of Health-Epidemiology Bureau. Additionally, modules allow laboratory data to be integrated and reconciled with local data. The PIDSR is a web-based tool, but the data is entered offline. The 2018 Joint External Evaluation noted *"Data management and validation is a lengthy process, done manually, to link laboratory and surveillance data and duplicate the records. At the time of the JEE, there were major reporting delays for measles with suboptimal completeness and timeliness of the overall reporting system"* (WHO 2018). For event-based surveillance, such as the COVID-19 outbreak, the Event-Based Surveillance and Response system, managed by the Epidemiology Bureau, allows all levels of government to report public health events. This system also allows data to be exchanged between local and national levels, as well as systematic monitoring of the media to take place. In 2018, the Joint External Evaluation noted that *"an international component has been implemented whereby cross border events of interest are included in reporting"* (WHO 2018) though the COVID-19 experience proved an ongoing need to enhance this component to more readily integrate with border health controls to monitor and manage imported cases.

73.   **Despite the need for enhancements and further integration of surveillance systems, current surveillance and reporting systems provide valuable data useful for epidemic and pandemic risk modeling.** The systems provide valuable data on both the severity and frequency of epidemic and pandemic events, including impact of age and comorbidity on disease progression and outcomes. Nonetheless, these events remain uncommon and unpredictable, and it is therefore a challenge to ascribe a loss value and develop insurance instruments to cover these. Global efforts are ongoing to use surveillance data to assess the role of risk reduction activities, population demographics, population density and comorbidities on the severity and spread of COVID-19 and, from this, the health insurance solvency implications. In this regard, the Philippines' current systems are fit-for-purpose, with data able to be used in modeling efforts.

74.   **Good systems are in place to track and publicize morbidity and mortality data**. The Department of Health holds such data across all disease areas, including data on infectious and novel diseases. The data is drawn from a number of electronic systems, and is supplemented by periodic surveys, which are typically patient or population facing. The data is published publicly in a timely manner through the Department of Health website and global databases the Philippines contributes to, such as the WHO's data collections. This allows public and private sector actors to develop reliable epidemiological models.

Of course, future novel diseases cannot be modeled in the same manner as diseases with known etiology, as transmissibility and severity are unknown. But data is sufficient on past epidemics and the recent pandemic to make well-informed multivariate models.

75.   **The Department of Health holds considerable data on overall health financing, including data on government planned and actual expenditure by program, health facility and disease-type**. The Department of Health and Department of Finance also hold and publicize data on government premium contributions to PhilHealth, from both national and local government unit budgets. However, PhilHealth expenditure data and the co-pays required by patients is less transparent. PhilHealth publicizes case-based payment rates and both PhilHealth and private insurers have clear benefit schedules. However, data on out-of-pocket co-pays is often lacking. Episodic level data on the level of co-pay paid by patients is rarely published. This makes it hard to track the total cost of patient episodes and may limit the development of new private health insurance products that close protection gaps. Historically transparency has been very limited on the cost of healthcare delivered by the private sector, complicating this issue. In 2021, the Medical Bill Transparency Act (Bill No. 2334) was introduced in the Senate to require all healthcare providers to provide transparent pricing data so that patients are fully aware of the total costs of their care, including any co-pays, and therefore face "no surprises." This parallels global efforts to enhance the transparency of provider pricing and tackle high rates of medical inflation. Although facing pushback from private providers, this proposed act is a positive development and will enhance data availability on healthcare costs, including those encountered during epidemics and pandemics. This will further inform reliable costed risk models that underpin product development.

3.1.3   Diagnostic and Recommended Actions

76.   **Work is ongoing to establish fixed asset registers in the government. The valuation of fixed assets is based on historic costs.** Identification and registration of individual fixed assets systematically is a prerequisite to insuring government assets. A well-maintained fixed asset register initiative is a key enabler of the development of risk transfer mechanisms as it can support the quantification of disaster risk and associated contingent liabilities, informing the design and pricing of sovereign and nonsovereign insurance programs and products. Recognition and registration of individual fixed assets is the fundamental common prerequisite of valuations for financial reporting, accounting and also for insurance purposes. Valuation is currently based on historic costs, but a revaluation exercises every 3 to 5 years would comply with accounting standards and would support the replacement cost approach to insurance.

> **Fixed asset registers, once established, should be reviewed every 3 to 5 years, including to update the valuation of assets if required. Valuation data should include replacement value.**

77.   **The Philippines grapples with two significant problems: a shortage of physicians and nurses and insufficient supply of resources across the country, which limits its capacity to respond to pandemics and epidemics.**

Strengthen financing of healthcare infrastructure to ensure adequate levels of health resilience to epidemics and pandemics, especially in the more remote areas of the country.

78.   **Historic and modeled risk data, aggregated and granular, to develop and design suitable DRF instruments is not readily available.** Considering that the Philippines is highly disaster prone, historic and modeled risk data are key in enhancing the financial management of disaster risk.

The government should continue to improve the availability of and access to disaster risk data, including ensuring wide availability of robust, granular data by government agencies, the insurance sector, and other nonsovereign users.

79.   **While important data collection and monitoring in case of pandemics and epidemics are in place, areas needing strengthening exist:**

The government, working with the WHO and other parties involved in the Joint External Evaluation process, should continue to evolve integrated health surveillance systems. *The inclusion of data from private health facilities is important to ensure comprehensive coverage of all potential points of patient presentation. The continued integration of indicator and event systems, and the use of automation for the reconciliation of case reports with laboratory results will be important to support real-time health surveillance data. Further development of the international cross-border surveillance module within the PIDSR would be beneficial for surveillance data. The government should continue to engage in regional and global health surveillance and security efforts, including post-event reviews to help inform future epidemic and pandemic risk models.*

The government should continue to make anonymized, aggregate data on infectious disease outbreaks publicly available. *The government already make considerable data on disease outbreak available. As the PIDSR is further integrated and evolved, and data produced at close to real-time, timely publication of this data would allow the insurance market to undertake more sophisticated pandemic and epidemic modeling, which may inform future product development.*

Data on the total cost of health delivery should be made available. *The implementation of Bill No. 2334, otherwise known as the Medical Bill Transparency Act, would help provide greater cost certainty to patients and insurers, as well as increase the reliability of data on the total cost of healthcare delivery. This would help ensure epidemiological models on epidemics and pandemics could be more readily costed, and, in turn, risk-based solutions explored.*

80.   **Agriculture as a key economic sector seriously affected by natural hazards requires special attention with respect to data availability.** In addition to hazard data, information on disaster damage and losses is also important to establish correlations between hazards and losses and develop risk models for agriculture.

The government can launch a dedicated portal for agriculture, where all this data can be integrated and made available to registered government as well as private sector entities. *The Philippines Statistics Authority is the nodal agency for publishing data on various sectors but a dedicated portal for agriculture data can boost focused applied research that can help the sector in many ways. This way the fragmented data sitting inside various government departments can be put to better use in the national interest. The combination of technology and data can be a game-changer for agriculture in the Philippines.*

81.   **Substantial investments in agritech will be needed to allow agriculture to withstand disasters efficiently and enhance the sector's insurability.** Essentially, agritech solutions can enhance farm level risk management, risk reduction, risk monitoring, yield estimation and loss assessment. Not just insurers but farmers, lending institutions, agri-input providers, traders, economists, climate scientists, policymakers and agriculture researchers can also benefit a lot out of agritech. Globally, many technological interventions for agriculture are being designed and implemented. Some of them are:

- *Agricultural robots and smart appliances*—Companies are developing and programming autonomous robots to handle essential agricultural tasks such as sowing seeds and harvesting crops at a higher volume and faster pace than human laborers. This is addressing the problem of availability, productivity, and high cost of agricultural labor.
- *Crop and soil monitoring*—Companies are leveraging computer vision and deep-learning algorithms to process data captured by drones and/or software-based technology to monitor crop and soil health. The outputs include advisories on optimal use of soil nutrients, fertilizers, and pesticides. These applications enable farmers to improve productivity.
- *Predictive analytics*—*Machine learning models* are being developed to track and predict various environmental impacts on crop yield such as weather changes. These analytics are used to alert farmers on adverse weather conditions well in advance, optimal sowing and harvesting periods, possible pest attacks, etc.,
- *Chatbots*—Specialized farming chatbots like Alexa are being developed to assist farmers in their day-to-day issues. Farmers can get customized advice from remotely located experts by sharing the ground level information through photographs on a range of issues.
- *Crop loss assessments*—These *assessments* through satellite imagery, drones and ground level data retrieved from mobile applications enable lending institutions and insurance companies to assess losses accurately and make timely payment of claims.
- *Market information and access through mobile applications.* These enable farmers to assess future commodity prices and thereby adjust their crop selection. Applications also link farmers to agriculture markets and potential buyers thus enabling them to get the best price for their produce without having to deal with intermediaries.

Develop agritech to increase efficiency and predictability and thus reduce agriculture risk to a level where insurance becomes sustainable and affordable.

3.2 Government Policy

3.2.1 Sovereign and Household Disaster Risk Financing Instruments

82.   **The 1951 Property Insurance Law requires public assets to be insured.**[54] Section 5 of the Property Insurance Law requires every government, except a municipal government below first class, to insure its properties at replacement cost, against losses due to fire, earthquake, storm, or other casualty. The insurance is to be taken from the Property Insurance Fund established under the same law and administered by the Government Service Insurance System (GSIS) against any insurable risk and pay the premiums thereon, which, however, shall not exceed the rates charged by private insurance companies. Notwithstanding the legal requirements to have public assets insured, a number of assets remain uninsured. Underinsurance is also an issue. As a response to the noncompliance with the law, on May 2018 a government circular[55] reinforced the requirement of insuring public assets. However, the noncompliance appeared to remain an issue at the time of writing of this report based on several indications to the consultant team (Appendix).

83.   **A parametric catastrophe risk insurance product to support the government's financing disaster losses was purchased in July 2017.** The product was placed with the World Bank who in turn transferred the risk to international reinsurers that in turn transferred the risk to "retrocessionnaires." The policy provided maximum coverage of ₱10.4 billion ($206 million) in the first year, and ₱20.5 billion ($406 million) in the second year, equally split across two components: (i) coverage for 25 provinces against emergency losses from major typhoons (with payouts based on modeled loss); and (ii) coverage for national government agencies against emergency losses from major typhoons and earthquakes for national government assets. The Bureau of the Treasury paid the premium, held the policy, and itself (with other relevant government departments) decided how to share the payout between affected provinces. Figure 7 details the product and its structure. The pilot was discontinued after 2 years with a loss ratio of 41.4%.

84.   **The Philippine disaster risk transfer instruments have included capital market solutions.**[56] The Philippines acquired World Bank-issued two-tranche catastrophe-linked bonds (CAT bonds IBRD CAT-123 -124) in 2019. The instruments had a 3-year duration and provided financial protection of up to $75 million for losses from earthquakes and $150 million for losses from typhoons (calibrated to losses of a category 5 typhoon). The recovery or risk trigger, like in the parametric catastrophe risk insurance, was a modeled loss resulting in payments at losses of 35%, 70%, and 100% of the sum at risk. The annual cost of the bonds had a minimum floor equal to a risk margin of 5.50% (earthquake) and 5.65% (typhoon) of the sum at risk. One payout of $52.5 million was made following Typhoon Rai (Odette) in December 2021. The bond matured without any further payout. As of late 2022, the government was not seeking a CAT bond renewal. Instead, the Bureau of the Treasury informed the consultant team that indemnity insurance of public assets was being prioritized. Treasury also noted the country's positive experience with contingent disaster financing from ADB, Japan International Cooperation Agency, and the World Bank (para. 42).

---

[54] Republic Act No 656: Property Insurance Law. Republic Act No. 656–Government Service Insurance System. gsis. gov.ph.

[55] Commission of Audit. Circular 2018–002. https://www.coa.gov.ph/wpfd_file/coa-circular-no-2018-002-may-31-2018/?hilite=2018–002.

[56] Information on these instruments is taken from Artemis: https://www.artemis.bm.

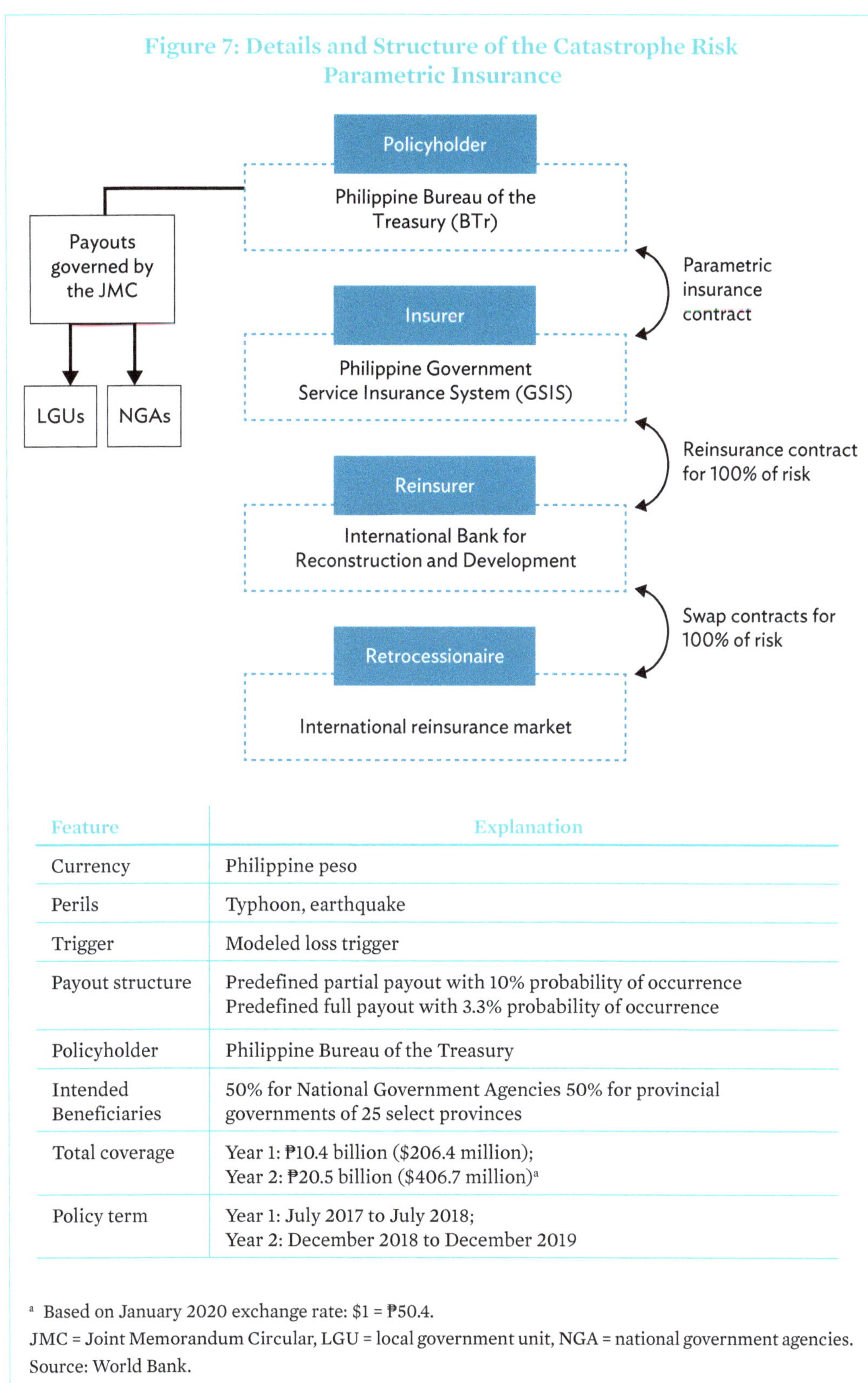

Figure 7: Details and Structure of the Catastrophe Risk Parametric Insurance

Feature	Explanation
Currency	Philippine peso
Perils	Typhoon, earthquake
Trigger	Modeled loss trigger
Payout structure	Predefined partial payout with 10% probability of occurrence Predefined full payout with 3.3% probability of occurrence
Policyholder	Philippine Bureau of the Treasury
Intended Beneficiaries	50% for National Government Agencies 50% for provincial governments of 25 select provinces
Total coverage	Year 1: ₱10.4 billion ($206.4 million); Year 2: ₱20.5 billion ($406.7 million)[a]
Policy term	Year 1: July 2017 to July 2018; Year 2: December 2018 to December 2019

[a] Based on January 2020 exchange rate: $1 = ₱50.4.

JMC = Joint Memorandum Circular, LGU = local government unit, NGA = national government agencies.

Source: World Bank.

85.   **A Philippine City Disaster Insurance Pool concept providing parametric insurance was developed in 2018.** Notwithstanding the existence of local special purpose disaster funds and mandatory indemnity cover for public assets (para. 83), cities often face significant challenges in securing adequate resources for post-disaster operations, including rapid access to funding to support early recovery efforts such as the restoration of critical infrastructure, delivery of services, and support of livelihoods. To address these needs, the Philippine City Disaster Insurance Pool was developed to directly support the second (local government) of the three tiers of DRF under the government's 2015 Disaster Risk Financing and Insurance Strategy (ADB 2018). The Philippine Department of Finance led the design of the pool, with technical assistance from ADB. The proposed triggers are spectral acceleration (a measurement of ground motion) to determine its payouts from earthquakes and a 3-second peak gust (a measurement of wind speed) to determine pool payouts from typhoons. These triggers are correlated with physical damage. The pool arrangement offers a number of benefits to participating cities including risk diversification, economies of scale, and knowledge sharing.

86.   **As a potential prelude to the introduction of this pool, GSIS is developing a city-level parametric disaster insurance product for earthquake and typhoon cover.** Under this scheme, GSIS will issue individual policies directly to each participating city rather than establishing a pool, with coverage intended to be available from 2023. The parametric triggers will be based on those designed for the Philippine City Disaster Insurance Pool. This scheme will be the first of its kind in Southeast Asia (ADB 2020).

3.2.2 Diagnostic and Recommended Actions

87.   **The use of the provincial parametric catastrophe risk insurance product provided important insights to the government.** A key purpose of the parametric catastrophic insurance product is to provide rapid liquidity after a disaster, since there is no need for the loss adjustment as the claim payment is based on a pre-defined trigger. In this case, that is based on modeled loss[57] resulting from the scale of event experienced. According to the World Bank's lessons learned report (World Bank 2020), this objective was not achieved, due to the delay in the transfer of the funds to the ultimate beneficiaries, the local government units and national government agencies. While the funds arrived within the specified timeframe to the Bureau of the Treasury, the process to assign the funds to the local government units and national government agencies was not effective; resulting in significant delays and disappointed expectations. In addition, the involvement of several parties resulted in high costs of coverage. Among the positive outcomes of the use of the product, however, was the capacity building within the government on the use of sophisticated disaster risk transfer instruments to the international markets, as subsequently demonstrated by the 2019 placement of a CAT bond in the capital markets.

> *The process to disburse claims payments to the ultimate beneficiaries should be timely and based on a pre-established methodology.*

---

[57] Modeled loss is one of the options used for parametric insurance. Other forms include the intensity of the event, like the wind speed or earthquake magnitude.

***The placement of parametric insurance should limit the number of involved
parties to the necessary minimum to contain intermediation costs.***

88.   **CAT bonds, as with insurance, carry potential basis risk.** Given the complexity
in accurately calculating the loss given the covered event using a probabilistic model,
the difference between the actual loss and the probabilistic modeled loss could be large.

***If the payout trigger is related to a model loss calculation, such a model should
be fully understood by the government including the ultimate beneficiaries to
determine the usefulness of the disaster financing instrument.***

89.   **There is lack of or under-insurance of assets critical to disaster responses.**
Notwithstanding the 1951 Property Insurance Law requiring public assets to be insured, assets
under local government units control are, however, generally uninsured or under-insured.
A lack of incentives and/or penalties for taking out insurance inhibits widespread uptake of
insurance critical to disaster response.

***Following the completion of a comprehensive register of assets, consideration
could be given to the development of an incentive scheme to encourage local
government units to take out appropriate insurance to public assets.***

3.2.3 Epidemics and Pandemics Risk Management and Financing Instruments

Epidemic and pandemic risk management

90.   **The Philippines has arrangements to allow for the cross-government management
of epidemics and pandemics.** The Inter-Agency Task Force for the Management of Emerging
Infectious Diseases was established through Executive Order No. 168 in 2014.[58] It was
conceived as the government's vehicle to manage and prevent the spread of any potential
emerging infectious disease outbreak. The composition includes all major departments
and units of government, including the departments of Finance, Justice, Education, Health,
Social Welfare and Development, as well as the Philippine National Police, the armed forces,
the coast guard, government-owned and controlled corporations, and local government
units. The taskforce chairmanship sits with the Department of Health, though, in practice,
the President of the Philippines chaired several meetings of the Inter-Agency Task Force
during the COVID-19 pandemic.

---

[58] Government of the Philippines, *Executive Order No 168 - Creating the Inter-Agency Task Force for the management
of emerging infectious disease in the Philippines,* 2014. Executive Order No. 168, s. 2014 | Official Gazette of the
Republic of the Philippines.

91.   **During the COVID-19 pandemic, the Inter-Agency Task Force for the Management of Emerging Infectious Diseases was convened and developed a cross-cutting action plan to manage the disease outbreak.** The task force was convened in January 2020 to address the growing concern over the outbreak of COVID-19 in Wuhan, People's Republic of China. This was in response to the declaration by the WHO of a global public health emergency. At this stage, much was unknown, particularly around the severity and contagiousness of the virus, and actions taken were limited. In March 2020, following the declaration of a pandemic event by the WHO, the President called the Inter-Agency Task Force for the Management of Emerging Infectious Diseases to coordinate the response in the Philippines. Two weeks later, the task force revealed a National Action Plan to manage the spread of COVID-19 in the community.[59] The plan was updated throughout the pandemic and is now in its fourth version, which focuses on opening up the country and rebuilding the economy. On 30 April 2020, through Executive Order no. 112, the President delegated authority for the imposition of community quarantine to the Inter-Agency Task Force for the Management of Emerging Infectious Diseases (effective 16 May 2020).

92.   **Due to the potential scale of the pandemic, the task force created the National Task Force Against COVID-19 pandemic chaired by the Secretary for the Department of National Defense.** The National Task Force oversaw the effective operationalizing of policy decisions made by the task force. Several other sub-taskforces were created with specific responsibilities, notably the COVID-19 Shield chaired by the head of police, which was charged with ensuring adherence to COVID-19 rules and laws.

93.   **Zoonotic disease management is underdeveloped, which is likely to make the risk of economic losses from epidemics and pandemics high.** In the Philippines, 75% of infectious agents and diseases over the last 30 years have been zoonoses (e.g., H1N1, SARS, HPAI, leptospirosis and rabies) (WHO 2018). In 2011, Executive Order Number 10 created the Inter-Agency Committee on Zoonoses (PhilCZ) as *"a collaborative mechanism.......to synergize and harness the strengths and capabilities of the concerned departments and bureaus of the government for efficient use of resources in the control and eventual elimination of existing zoonoses."*[60] The PhilCZ's central objective is to *"endeavor to: a) develop a national strategy on prevention, control and elimination of zoonoses; and b) establish a functional and sustainable mechanism to strengthen the animal-human interface for the effective prevention, control and elimination of zoonotic diseases."* PhilCZ, formed by the Department of Health, the Department of Agriculture, and the Department of Environment and Natural Resources works on five priority diseases: rabies, anthrax, leptospirosis, Japanese B Encephalitis, and Ebola Reston. Joint teams also collaborate at central and regional levels, and border points of entry into the country, to manage the risks of outbreaks and spread. However, beyond these areas, PhilCZ has not evolved significant plans and strategies for managing zoonotic disease and does not have a plan for the management of an epidemic or pandemic outbreak arising from a novel pathogen.

---

[59] Government of the Philippines, National Taskforce against COVID-19, *National Action Plan against COVID-19,* 2020 with updates in 2021 and 2022.
[60] Republic of the Philippines, Office of the President, *Executive Order 10,* 2011.

94.   **The Philippines lacks integrated and comprehensive preparedness and response plans for pandemics and epidemics.** The most recent WHO Joint External Evaluation processes revealed that *"there are several plans and manuals that guide emergency preparedness in the Philippines—the overarching National Disaster Preparedness Plan 2015–2028; Disaster Preparedness Manual for guidance in the local level; multi-hazard response plans; Manual of Operations and SOPs for Health Emergency and Disaster Response Management; the Ninoy Aquino RESPOND 39 of International Health Regulations Core Capacities of the Republic of the Philippines International Airport Public Health Emergency Contingency Plan 3rd Edition (2018); the Philippine Avian Influenza Protection Program; as well as implementation plans for major gatherings. Chemical, biological, radiological and nuclear (CBRN) simulation exercises, including the field management of mass casualties during a chemical incident was undertaken in 2017"* (WHO 2018). However, these plans lack sufficient integration and have significant gaps in relation to preparedness for, and response to, pandemics and epidemics. The Joint External Evaluation noted the need to *"Develop a generic emergency plan for communicable diseases that can become public health emergencies, such as pandemic influenza"* (WHO 2018). The need for health service resource mapping and identification of the role and available finance for pandemics and epidemics at local government unit level were also noted as areas for further work.

95.   **The COVID-19 outbreak uncovered fragile resilience against epidemics and pandemics.** The challenges are reflected by a global study on the resilience and response of economies to the coronavirus pandemic. The Philippines was ranked 48 out of the 53 economies studied in the Bloomberg Covid Resilience Ranking as of April 2022 (Bloomberg 2022), with a score of 55.5, behind Viet Nam (66.7), Indonesia (66.3), Malaysia (68), and Thailand (63.7). Norway's score was the highest (83.1). The Covid Resilience Ranking is a monthly snapshot of where the virus is being handled most effectively with the least social and economic upheaval. Drawing on 12 data indicators spanning virus containment, quality of healthcare, vaccination coverage, overall mortality and progress toward restarting travel, it captures how the world's biggest 53 economies have responded to the same once-in-a-generation threat.

96.   **In response to the recent COVID-19 outbreak, plans were announced in 2021 to create the Center for Disease Control and Prevention (CDC).** The limited staffing of the Health Emergency Management Bureau means its work and responses have largely been to disasters triggered by natural hazards and climate change, with minimal work on epidemic and pandemic health preparedness or management. The Health Emergency Management Bureau works with Disease Prevention and Control Bureau (DCPB) and the emergency department, but nonetheless faced organizational and resource issues during the COVID-19 pandemic, which perhaps limited its effectiveness. Consequently DCPB proposed that the CDC be created as the country's technical authority on health security and preparedness. Senate Bill 2505 seeks to establish the CDC under the Department of Health as the technical authority on forecasting, preventing, controlling, and monitoring communicable and non-communicable diseases (Republic of the Philippines 2022). It is proposed that the CDC absorb units such as the Epidemiology Bureau, the Research Institute for Tropical Medicine, and other relevant functions. However, it remains to be seen how effective this will be at monitoring and managing future pandemics and epidemics, particularly because resources to the new entity effectively have not been costed or identified.

97.   **In most cases, zoonotic disease, whether bacterial, viral, or fungal in nature, spreads to people through contact with animals carrying the disease.** Having clear plans to tackle the illegal wildlife trade and manage animal-human interactions is necessary to reduce the risk environment and likelihood of contagion. Being home to a significant number of endemic wildlife species, the Philippines has measures to ban the illegal trade of wildlife, which also helps to prevent disease transmissions and outbreaks. In recent years, the government has increased enforcement capacity and efforts and made several significant seizures. However, given that wildlife trafficking is estimated to be the fourth largest transactional crime globally, there is likely significant levels of ongoing risk. The 10-year national Wildlife Law Enforcement Action Plan 2018–2028, which is aligned with the Philippine Biodiversity Strategy and Action Plan, serves as the national road map to address wildlife crimes and a guide to prioritizing enforcement activities, allocating funds and resources, and evaluating impacts of enforcement.[61] However, the demand for, and sale of, illegal wildlife in the Philippines is also closely related to cultural norms and expectations of citizens. A number of species, in which potential zoonotic disease is high, such as pangolins, continue to be in-demand—less for subsistence reasons and more for their use in traditional medicine.

98.   **Like many countries, the Philippines does not have a robust approach to the stockpiling of essential medicine and equipment in preparation for emergencies.** In April 2021, a Health Procurement and Stockpiling Act[62] was introduced to the House of Representatives, which sought to create a health procurement and stockpiling bureau under the Department of Health, financed by a new Medical Stockpiling Fund. This newly created bureau would be charged with procuring goods—such as personal protective equipment (PPE) and basic medicines—as part of a wider approach to readying for future healthcare emergencies. However, this proposed act is yet to be passed. The Department of Finance has expressed concerns over the Department of Health's ability to manage the fund and suggested this be handled by the Department of Finance. There are also questions on how a health stockpiling function would work in the context of increasing decentralization of health financing and system strengthening functions from national to local government contexts. Globally there is little consensus on the most effective arrangement to ensure adequate stocks of health consumables, equipment and medicines are available during epidemics and pandemics. Many countries do prepare by stockpiling certain items, such as PPE, either at national or regional levels. However, given the infrequency of epidemic/pandemic events, and the finite shelf life of many health products, there are other countries which consider stockpiling to be wasteful. The main consideration for policymakers is to ensure that if stockpiling isn't pursued, that there are arrangements in place to allow in-demand items to be purchased in the event of simultaneous global demand, such as a contract with a distributor/manufacturer to supply the country preferentially and at pre-agreed rates.

---

[61] Department of Environment and Natural Resources. Administrative Order No. 2020–13: Wildlife law enforcement action plan 2018–2028, 2018 DAO-2020-13 (emb.gov.ph).

[62] Government of the Philippines, *House Bill 6995: An act providing for the stockpiling of strategic and critical drugs and medicines, vaccines, devices and materials for public health emergencies, creating for the purpose the health procurement and stockpiling bureau under the Department of Health funds thereof.* HB06995.pdf (amazonaws.com).

The health system structure and regulation

99.   **The Philippines has made steady progress toward universal health coverage without crowding out the private sector.** Health as a basic human right is enshrined in the 1987 Philippine Constitution (Article II, Section 15), which declares "the State shall protect and promote the right to health of the people and instill health consciousness among them."[63] Under this mandate, the Department of Health, as the national technical authority on health, has the responsibility to ensure the highest achievable standards of health care, from which local government units, nongovernment organizations, the private sector and other stakeholders anchor their health programs and strategies.[64] In 2019, the Government of the Philippines signed into law the Republic Act No. 11223, commonly referred to as the Universal Health Care Act,[65] which further aimed to reform and extend the social health system and provide equality of access and quality of care for all Filipinos. The National Health Insurance Programme is carried out by PhilHealth, the state's social health insurance provider. This is financed by both mandatory individual and employer contributions and government subsidies (Figure 8 and Table 8). The premiums paid to PhilHealth are low by international standards, at just 3% of income for formal workers, but this is set to rise gradually to 5% by 2025. Historically, low premiums have resulted in PhilHealth's benefits being skewed toward protection for inpatient services and catastrophic claims for most members. However, reforms brought about by the Universal Health Care Act now mean primary care provision is covered for members.[66] It is perhaps too early to assess the impact benefit expansion will have on scheme sustainability, but strengthening primary care is widely considered beneficial in tackling high out-of-pocket payments (Figure 9) and coping with the rise in non-communicable diseases. PhilHealth has moved to case-based payments for most procedures and clinical activities. If costs are incurred above these case rates, members bear liability for these. Most public health facilities try to adhere to the case rate, but private providers often charge considerably more. This has spurred the growth of the private health insurance sector, which comprises both health maintenance organizations (HMOs) and private insurers. Plans offered by these firms usually work in conjunction with PhilHealth benefits, and help close out-of-pocket protection gaps while extending access to a wider range of providers (section 3.3.7).

100.   **The Philippines has a complex health financing environment with the Department of Health, PhilHealth, local government units and private insurers all contributing to the financing of premiums and services.** The Local Government Code of 1991 devolved the authority, responsibility and financing of certain health services to different tiers of government. This includes devolution of public health programs, management of local government hospitals, preventative health and health promotion activities to local government units.[67] A range of sources also contributes to the national health insurance scheme, managed by PhilHealth, including local government units, national government, and members. PhilHealth accounts for 41.2% of public spending on health—including general government spending and compulsory health insurance—while the national Department of Health and local government units control 38.8% and 20%, respectively.

---

[63] The Constitution of the Republic of the Philippines, 1987, Section 15. For information, see The Constitution of the Republic of the Philippines | Official Gazette of the Republic of the Philippines.

[64] The Republic of the Philippines, Office of the President, *Executive Order 102*, 1999.

[65] The Republic of the Philippines, *Republic Act 11223: An Act instituting universal health care for all Filipinos, prescribing reforms in the health care system, funds thereof*. 2019.

[66] 3 PhilHealth Circular 2020–0020 Governing Policies of PhilHealth Konsultasyong Sulit at Tama (PhilHealth Konsulta) Package: Expansion of the Primary Care Benefit to Cover All Filipinos.

[67] Republic Act 7160: An act providing for a local government code, 1991.

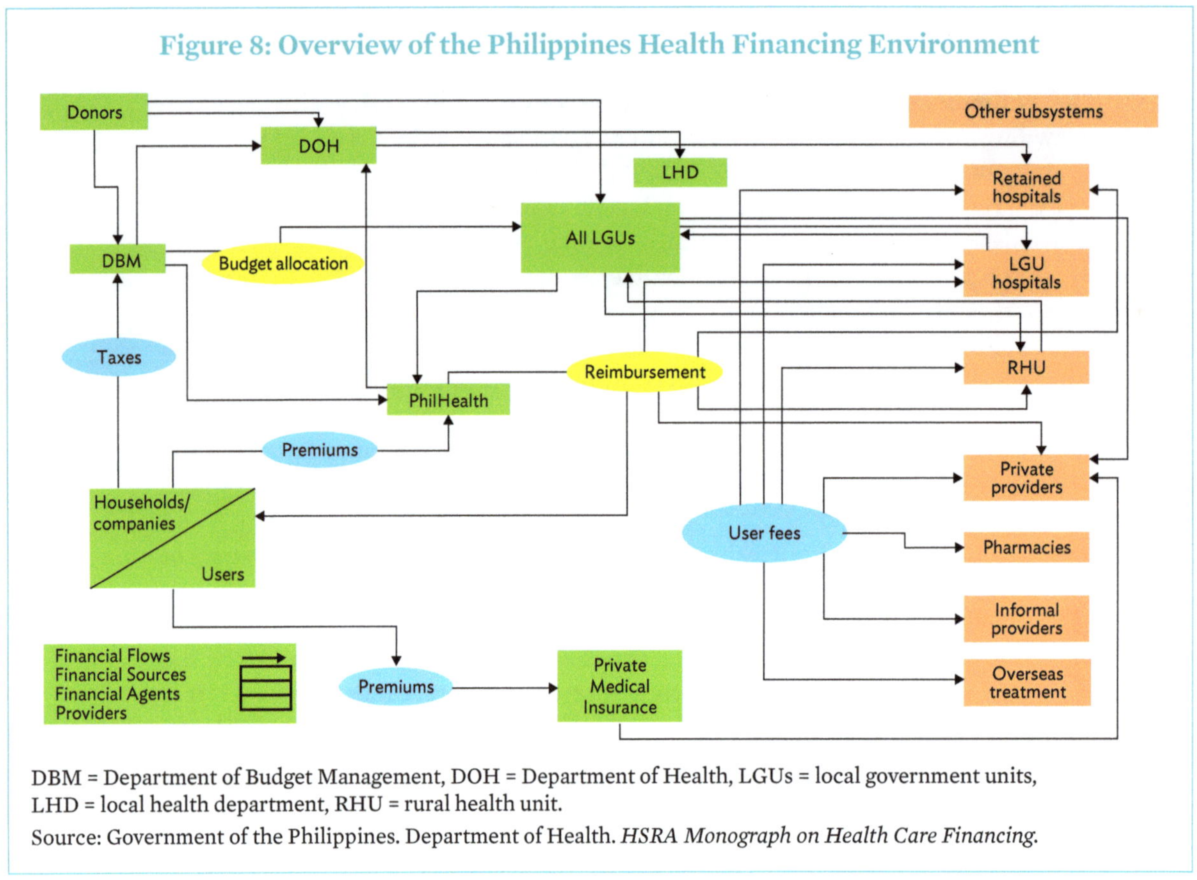

Figure 8: Overview of the Philippines Health Financing Environment

DBM = Department of Budget Management, DOH = Department of Health, LGUs = local government units, LHD = local health department, RHU = rural health unit.

Source: Government of the Philippines. Department of Health. *HSRA Monograph on Health Care Financing*.

Table 8: PhilHealth Membership Groups, Premiums, and Financing Source

Group	Premium	Paid by	Eligibility
Employees with formal employment	3.5% of income (rising by 0.5% annually to reach a maximum of 5% by 2025)	Split equally between employer and employee	All workers in the formal sector
Self-earning individuals; Professional practitioners	3.5%–4% of income (rising by 0.5% to reach a maximum of 5% by 2025) The current income floor and income ceiling for voluntary members are also fixed at ₱10,000 and ₱80,000, respectively. This income ceiling will likewise increase by ₱10,000 every year until it reaches ₱100,000 in 2024/2025.	Individuals	Those who work for themselves and are therefore both the employer and employee are qualified under this program

continued on next page

Table 8   *continued*

Group	Premium	Paid by	Eligibility
Kasambahay (domestic workers)	*Kasambahay* have the same PhilHealth contribution rate and computation as formally employed members.	For household workers receiving a monthly salary of ₱5,000 (and below), their employers are required to pay their total monthly contribution in full to PhilHealth. However, *kasambahay* earning more than ₱5,000 should share half of their total monthly contribution payment.	Domestic helpers working in the Philippines
Overseas Filipino Workers	4% of Monthly Basic Salary	Individuals	Migrant workers
Foreigners working in the Philippines	Retirees pay a flat rate of ₱15,000. Expats exchange students, and other foreigners pay a flat rate currently set at ₱17,000. These are the highest rates among all categories.	Individuals	Foreigners living, working, or retiring in the Philippines
Lifetime members (definition under eligibility column)	Retirees with at least 120 contribution payments who are registered as lifetime members no longer need to remit to PhilHealth. However, lifetime members who become employees in the Philippines or abroad must resume making PhilHealth contribution payments until they resign or get terminated from work	…	• Individuals aged 60 years and above and have paid at least 120 monthly contributions with PhilHealth and the former Medicare Programs of Social Security System and GSIS; • Uniformed personnel aged 56 years and above and have paid at least 120 monthly contributions with PhilHealth and the former Medicare Programs of SSS and GSIS; • SSS underground miner-retirees aged 55 years above and have paid at least 120 million monthly contributions with PhilHealth and the former Medicare Programs of SSS and GSIS; • SSS and GSIS pensioners prior March 4, 1999

continued on next page

Table 8   *continued*

Group	Premium	Paid by	Eligibility
Indirect contributors including: Indigents identified by the Department of Social Welfare and Development Beneficiaries of the Pantawid Pamilyang Pilipino Program, senior citizens, people with disabilities, Sangguniang Kabataan officials previously identified at point-of-service/sponsored by local government units	Contributions are in line with the rates for employed individuals (currently 4% and rising to 5% by 2025)	Government either nationally or through local government units If any individual falling within any "indirect contributor" group takes on employment and meets the contribution thresholds then contributions are made in accordance with the provisions set out in the rest of this table.	Each indirect contributory group has different criteria. But these categories are generally reserved for vulnerable individuals. This includes families identified as ID-poor, those identified as having a diagnosable disability and senior citizens who do not meet the criteria for lifetime membership

... = not available, GSIS = Government Service Insurance System, PhilHealth = Philippine Health Insurance Corporation, SSS = Social Security System.

Source: PhilHealth website, 2022. https://www.philhealth.gov.ph/services/.

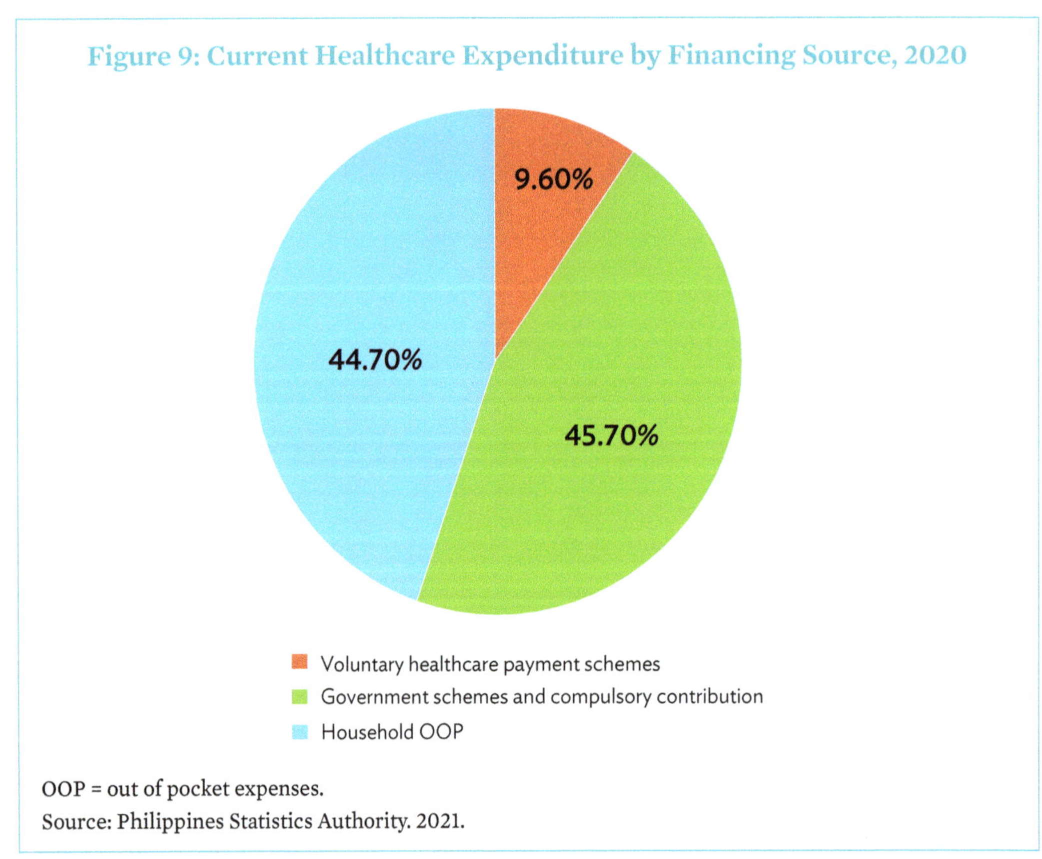

OOP = out of pocket expenses.
Source: Philippines Statistics Authority. 2021.

101.   **There is a lack of clarity on PhilHealth coverage for novel infectious diseases, possibly resulting in an epidemic and pandemic health protection gap**. It is common for social insurers to exclude coverage for prevention, treatment, and immunization costs arising from pandemic and epidemics, due to a lack of epidemiological and cost data to estimate the severity and frequency of such events. During the COVID-19 pandemic, PhilHealth was directed to extend benefits as a means of closing the protection gap. This served the dual purpose of ensuring access to healthcare and, as a consequence, increasing surveillance data on incidence and transmissibility.

102.   **During the COVID-19 outbreak, no additional funds were allocated for PhilHealth and coverage was given using its existing funds reserves.** At the outset of the pandemic, there was no clarity on how COVID-19 related healthcare needs would be financed. Under normal circumstances, PhilHealth, local government units, the Department of Health, and private insurers all play a role in financing healthcare. As noted in the 2018 WHO health system review *"Parallel funding by three sources* (Department of Health, PhilHealth and local government unit) *and lack of demarcation and harmonization in premium-funded benefits versus tax funded services are the primary reasons for confusion and inefficiencies in Philippine health-care financing"* (WHO 2018). This unresolved issue was exacerbated by the severity of the pandemic which delayed much needed financing reaching providers for whom the costs of providing COVID-19 care was unexpected and significant. The government debated providing additional financing to PhilHealth but ultimately decided it could not justify this given the organization's large reserve fund of ₱110 billion. PhilHealth was directed by the government to pay claims for COVID-19 from this reserve, that together with the government postponing the planned increases in premiums which were due to come into force to pay for new and extended benefit under the Universal Health Care Act, PhilHealth therefore needed to shoulder both the cost of COVID-19 treatment and the lost revenue from deferred increases in premiums. It is therefore perhaps unsurprising that the benefits package provided to policyholders to cover COVID-19 was often insufficient to meet the total cost of claims. Initially, factors such as the unknown course of the pandemic, the investigational nature of the drugs used, and the individually tailored clinical pathway contributed to unregulated processes and financial charges for the treatment of COVID-19. These costs often went above set benefit limits and contributed to high out-of-pocket payments. The government did, however, direct additional funds to other parts of the health system which are discussed further in this report.

103.   **From April 2020, PhilHealth introduced case-based severity rates with the intent of rationalizing benefits and standardizing the cost of COVID-19 treatment, but out-of-pocket payments remained high.** PhilHealth estimated the costs based on case rates for pneumonia given the parallels as potentially serious respiratory diseases. This is a reasonable approach given that little is known of the actual financial cost of unfamiliar diseases at the outset of a pandemic or epidemic; and thus, it is understandable for health systems to use estimates based on empirical data for similar diseases. A retrospective study of hospital expenditure of COVID-19 patients identified that despite adjustments, out-of-pocket payments remained high at 12% of total hospital expenditure, which is substantial and cost-prohibitive relative to personal incomes (Figure 10 and Table 9). The potential health care costs are likely to have created barriers for most people, potentially discouraging timely medical services and extending the cycle of infections.

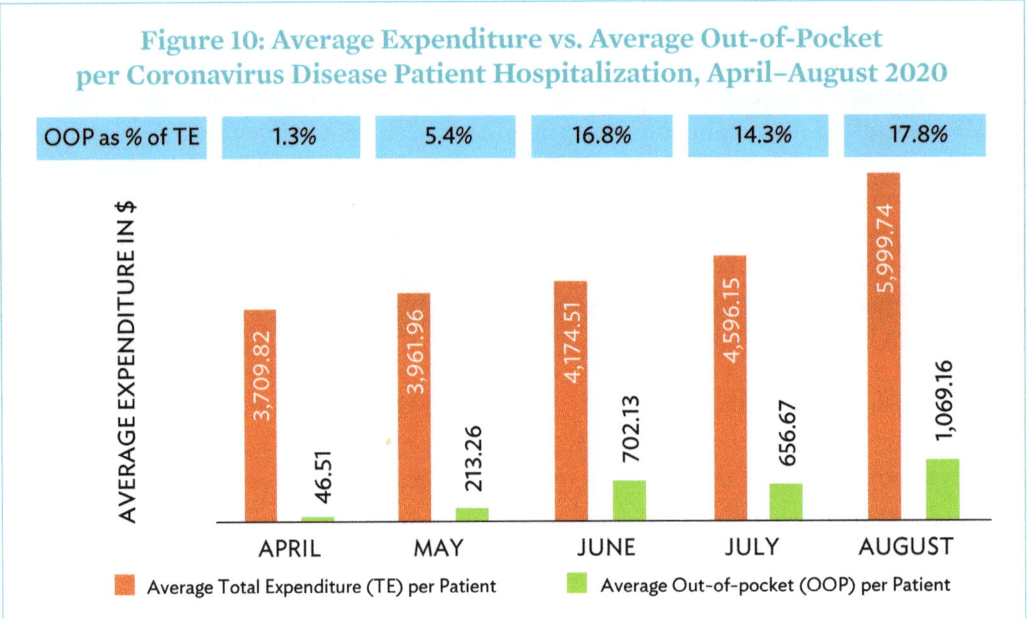

Source: Tabuñar, S.M. and Dominado,T.M. 2021. Hospitalization Expenditure of COVID-19 Patients at the University of the Philippines-Philippine General Hospital (UP-PGH) with PhilHealth Coverage. *Acta Medica Philippina*. Vol. 55 (2). https://actamedicaphilippina.upm.edu.ph/index.php/acta/article/view/2809.

Table 9: Coronavirus Disease Benefit Package and Case Rates Offered by PhilHealth

Code	Description	Case/Package Rate (in Philippine Peso)
SARS-CoV-2 Testing		
C19T1	All services and supplies for the testing are provided by the testing laboratory	₱8,150
C19T2	Test kits are donated to the testing laboratory	₱5,450
C19T3	Test kits are donated to the testing laboratory, cost of running the laboratory and the RT-PCR machine for testing are subsidized by the government	₱2,710
Community Isolation		
C19CI	COVID-19 Community Isolation Package	₱2,449
Z29.0	Admission to Protect the Individual from surroundings or for isolation of individual after contact with infectious disease	₱14,000
Inpatient Management/Care		
C19IP1	Mild pneumonia in older people or with comorbidities	₱43,997
C19IP2	Moderate pneumonia	₱143,267
C19IP3	Severe pneumonia	₱333,519
C19IP4	Critical pneumonia	₱786,384
COVID-19 Vaccine Injury Package		
C19VIH	Hospitalization due to Serious Adverse Effects (SAEs) following COVID-19 immunization	₱100,000 (max)
C19VID	Death or permanent disability due to Serious Adverse Effects (SAEs) following COVID-19 immunization	₱100,000 (lumpsum)

Source: National Health Insurance Program, Audit Report 2020–21 Philippines and PhilHealth Circular 2021, Annex A: Package Rates and Benefit Cap for the COVID-19 Vaccine Injury Package.

104.   **PhilHealth created a number of mechanisms for the disbursement of COVID-19 funds to healthcare institutions, but major challenges were noted by the Commission of Audit in how these were used.** PhilHealth has, as part of its operating procedures, an interim reimbursement mechanism which allows for advance payments to be made to healthcare institutions under exceptional circumstances, but only with the approval of the President of the Philippines. Early in the pandemic, PhilHealth utilized the interim reimbursement mechanism (IRM) to provide healthcare institutions with advance payments for anticipated COVID-19 cases. Hospitals could claim the equivalent of 3 months' worth of claims based on historical data, which was then charged against their future claims. However Presidential approval for the IRM was not sought or provided and the Commission of Audit noted *"PhilHealth released funds in the total amount of ₱14,971 billion to various healthcare institutions nationwide under the IRM for services not yet rendered, contrary to the prohibition against advance payments on government contracts under Section 88(1) of PD No.1445"* (Republic of the Philippines 2020, 7). In addition, various irregularities were found in the calculation of IRM payments, including overpayments and inappropriate handling of withholding tax. The Commission of Audit concluded that PhilHealth had not exercised proper diligence nor followed operational procedures and as a result, *"the disbursements made under the IRM scheme were without legal authority and could be considered illegal expenditures."*[68]

105.   **In April 2021, PhilHealth, under pressure from healthcare institutions to speed up payments for COVID-19 claims, worked with the Inter-Agency Task Force to create a new form of disbursement entitled the debit–credit payment method**. The method facilitated the settlement of accounts payable to healthcare facilities in areas identified by the Inter-Agency Task Force as high or critical risk for COVID-19. The debit–credit payment method allowed payment of 60% of the total applicable healthcare facility claims payable, with the remaining 40% of the total amount of claims paid following full compliance with existing claims processing requirements and procedures, and full reconciliation of the 60% of the total amount of receivables initially paid to the facilities. However, its slow implementation meant that a number of hospitals cut ties with the state insurer over high levels of accumulated unpaid claims, defined as claims 60 days aged or longer. This created access challenges for some patients, and created credibility issues for the country's state insurer.

106.   **PhilHealth operates within a complex system of decentralization, which means funds for frontline providers are often routed via local government units**. The special health fund was created by the Universal Health Care (UHC) law with the aim of supporting decentralization and appropriate local integration of funding. The UHC act provides that *"all health resources intended for health services to finance population-based and individual-based health services, health system operating costs, capital investments, and remuneration of additional health workers and incentives for all health workers shall be pooled to the Special Health Fund."*[69] However, it is unclear if the Special Health Fund (SHF) will be the vehicle for handling emergency funds that may be allocated during emergencies. During the COVID-19 pandemic, national public hospitals and providers had claims reimbursed by PhilHealth directly. However, providers under the management of local government units had their

---

[68] Republic of the Philippines, Commission of Audit, *Audit Report of the Philippine Health Insurance Corporation*, 2020, p. 7.

[69] Department of Health, Philippines, Memorandum Circular entitled DOH-DBM-DOF-DILG-PHIC Joint Memorandum Circular No. 2021–0001 entitled "Guidelines on the Allocation, Utilization, and Monitoring of, and Accountability for, the Special Health Fund," January 2021. See Scanned Image (doh.gov.ph), p. 2.

claims reimbursed via the local government units. There may therefore be some benefit in clarifying the approach to disbursements of emergency health fundings, including the reimbursement mechanism for all providers.

107.   **The Department of Health were provided with significant additional funds to support aspects of the COVID-19 response. The main mechanism to facilitate this was through adjustments to the national budget.** The funds were not specifically ringfenced for particular purposes so could be used across a range of activities including public health campaigns; procurement of vaccines and PPE; enhancing public health services at airports, ports, and other points of entry; and so on. In its most recent audit report, the Commission of Audit noted *"Various deficiencies involving some ₱67,323,186,570.57 worth of public funds intended for national efforts of combating the unprecedented scale of the COVID-19 crisis. These deficiencies contributed to the challenges encountered and missed opportunities by the Department of Health during the time of state of calamity/national emergency, and casted doubts on the regularity of related transactions"* (Department of Health 2020, p. 7). Moreover, the report noted major issues in the use of funds, noting contracts being let without proper documentation and process and the use of emergency funds for items that did not fit the legal basis for the funds, such as gift cards for staff. Most health systems grappled with similar issues, but the extent of the issues identified by the Commission of Audit were extensive. The Department of Health's normal role within the health system is as the countries' technical authority. Its functions are typically focused on policy and programming, rather than on financial management and procurement. As such, the Department of Health likely struggled to manage a considerable increase in public funding brought about by the COVID-19 pandemic. To its credit, the Department of Health publicly advised that the majority of issues noted in the 2020 Commission of Audit report were resolved (August 2021). It also contested some of the findings including those related to inappropriate fund transfers. The Department of Health stated that *"The ₱42.4 billion fund transfers were mostly to the Procurement Service (PS) of the Department of Budget and Management and Philippine International Trading Corp (PITC) for the procurement of COVID-19 supplies. COA observed that the funds were transferred without a Memorandum of Agreement (MOA), and despite delays in delivery of the procured medical equipment. The DOH responded that there is no need for a MOA for the transaction, since the items procured — PPEs, testing kits, and other equipment needed for the COVID-19 response have been classified as common-use supplies due to the pandemic."* (Department of Health 2020).

108.   **Mortality was higher in the Philippines than in many lower-middle-income countries in the region, some of which have poorer health systems and a less skilled healthcare workforce** (Figure 11). Several nonfinancial factors are likely to have contributed to this situation including poor levels of health literacy among the population, ongoing issues with basic sanitation and health behaviors and challenges in engaging vulnerable groups. However, a lack of centralized health emergency planning for pandemics and epidemics and clear definition of how the newly injected funds were to be used are likely to have been major issues.

109.   **The cost of COVID-19 vaccinations and other prophylactic interventions was significant**. The costs of vaccination programs extend to the cost of preventative healthcare activities, such as the delivery of major public health campaigns, the implementation resources as well as the vaccines themselves. The cost of vaccines, in particular, can be significant during

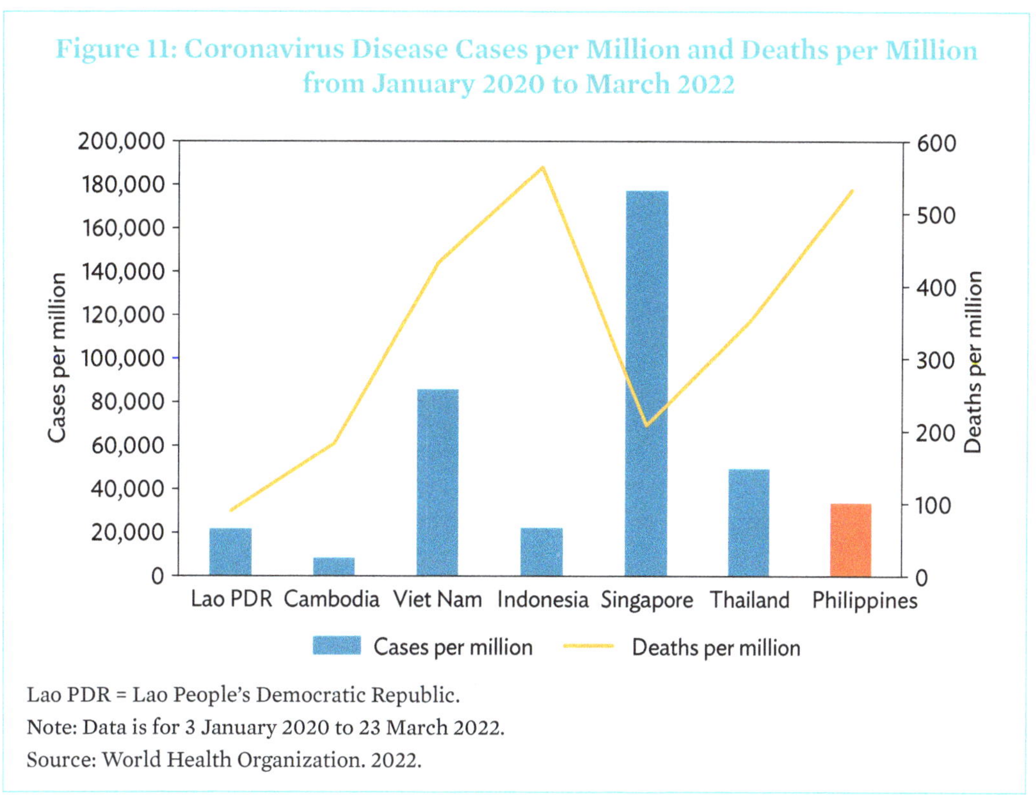

Lao PDR = Lao People's Democratic Republic.
Note: Data is for 3 January 2020 to 23 March 2022.
Source: World Health Organization. 2022.

epidemics and pandemics due to simultaneous high global demand, compared with limited supply during. Emergency-use and rapid loans were pivotal in supporting the Philippines to procure vaccines effectively.

110. **In response to the COVID-19 pandemic, the Department of Finance has raised $22.55 billion in budgetary support financing from ADB, World Bank, Asian Infrastructure Investment Bank, Agence Franç aise de Dé veloppement, Japan International Cooperation Agency, Export–Import Bank of Korea–Korea Economic Development Cooperation Fund (KEXIM-EDCF), and foreign currency denominated global bonds**. Of this, a total of $1.5 billon was specifically to finance vaccine procurement, rollout and public health preparedness strengthening, with an additional $600 million for general health system strengthening and resilience. In addition, it is likely that some of the funds raised by the Department of Finance for general expenditure arising from the COVID-19 pandemic were also directed toward the direct health response.

3.2.4 Diagnostic and Recommended Actions

111. **Although a social state insurer, it is nonetheless important that PhilHealth's role in financing healthcare costs arising from epidemics, pandemics or other classes of disasters, be made clear.** This helps ensure the insurer has appropriate controls and approaches to the management of its premiums and reserves. While PhilHealth was able to provide policyholder benefits for COVID-19 as the result of considerable reserves, this may not always be the case and is therefore an unreliable method of approaching disaster risk financing.

There should be clear articulation of PhilHealth's role in financing the direct healthcare costs arising from epidemics and pandemics.

112.   **Reinsurance of PhilHealth could be explored to help mitigate the government's exposure to direct healthcare costs during an epidemic or pandemic.** Reinsurance would allow for either a category of risk (such as pandemic risk) or a proportion of claim risk to be transferred out of PhilHealth's portfolio. Reinsurance options are commercially available to limit pandemic and epidemic risk. The most common form of reinsurance for epidemics and pandemics is stop-loss reinsurance. This is a form of nonproportional reinsurance in which the reinsurer makes a payment if the claim ratio exceeds a certain threshold. This could, therefore, protect PhilHealth from significant and sudden increases in claims that often occur during epidemic and pandemic events. There are also commercial reinsurance options based on quota share, a form of proportional reinsurance, where the reinsurer assumes a specific percentage of every risk being reinsured. In this model, PhilHealth and the reinsurer would share proportionally in all premiums and losses from a pandemic or epidemic event. In addition, reinsurance brings added advantages, like pandemic risk models capabilities, risk based pricing, capital relief, etc.

Reinsurance as a means to protect PhilHealth in case of a pandemic or epidemic should be explored. If found to be available and economically beneficial, the acquisition of such a protection should be pursued.

113.   **Provider payment mechanisms that PhilHealth can use during disasters need clarification.** A large single payer health system offers significant benefit at times of disaster as it is able to channel funds to healthcare providers quickly and through a unified mechanism. The IRM and debit-credit payment method mechanisms that PhilHealth have developed are promising, but oversight of their use needs strengthening. In addition, while decentralization of routine health financing to local government unit's is in line with current policies, there may be benefit in exploring alternate arrangements for emergency funds released at times of national disaster. This could include channeling emergency funds for hospitals and providers under local government unit control directly to the providers rather than via local government units to ensure funds reach frontline entities more quickly, in a more equitable manner and with greater transparency. Such a system would need careful exploration to ensure this would undermine local government unit ability to discharge their role in managing health services more broadly.

Reassess the provider payment mechanisms that PhilHealth can use during disasters.

114.   **Health security and preparedness need strengthening in line with international health regulations standards and building on the recommendations of the WHO Joint External Evaluation in 2018.** Of particular importance is the creation of an integrated cross-government strategy for the management of epidemics and pandemics, aligned with a national integrated policy for countermeasures deployment.

The strategy to strengthen health security and preparedness should include:

- *A scheme of delegated authority and accountability for managing disease outbreaks between different tiers of government and different government departments.*

- *The approach that will be taken to finance medical countermeasures, including the ex-ante financing of appropriate stockpiles and preparedness measures and ex-post financing of needs that emerge during the health event. This should also explain procurement governance, with clear roles and responsibilities for procurement under emergency situations, including disasters, pandemics, and epidemics. Aligning with the findings of the Joint External Evaluation, this should also "Clearly identify processes to fast track procurement, fund mobilization for procurement and registration for use of emergency vaccines."[70]*

115.   **The COVID-19 pandemic has led to delays in planned UHC reforms, including the extension of outpatient benefits to formal workers.** There is a need to refocus efforts on UHC activity to help close protection gaps and ensure comprehensive coverage is in place, particularly in relation to primary care. The current and planned ADB loans supporting UHC, and primary care development in particular, are important and will help ensure better health system resilience. Primary healthcare coverage for the population will help reduce out of pocket payments, encourage access, and reduce the likelihood of avoidable spread of disease.

> **Maintaining progression toward comprehensive UHC that reduces existing health protection gaps, bolsters health spending and enhances overall system resilience should be pursued.**

116.   **Pandemic and epidemic risk reduction efforts should be enhanced to encourage economically viable insurance to emerge.** The availability of insurance products covering both the health and business costs associated with pandemics and epidemics are highly affected by the risk reduction environment. For pandemics and epidemics, this is closely linked to a country's capability to manage communicable disease generally, and therefore limit the risks of contagion.

> **In this regard, the Philippines would benefit from making further progress toward full compliance with the International Health Regulations standards, utilizing the findings of the 2018 JEE (Joint External Evaluation).**

117.   **Zoonotic disease management is underdeveloped, which is likely to make the risk of economic losses from epidemics and pandemics high, as noted earlier.** In the Philippines, 75% of infectious agents and diseases that have emerged over the last 30 years have been zoonoses (e.g., H1N1, SARS, HPAI, leptospirosis and rabies) (WHO 2018). Again reiterating, in 2011, Executive Order Number 10 created the Inter-Agency Committee on Zoonoses (PhilCZ) as *"a collaborative mechanism…….to synergize and harness the strengths and capabilities of the concerned departments and bureaus of the government for efficient use of resources in the control and eventual elimination of existing zoonoses."[71]* The PhilCZ's central objective is to *"endeavor to: a) develop a national strategy on prevention, control and elimination of zoonoses; and b) establish a functional and sustainable mechanism to strengthen the animal-human interface for the effective prevention, control and elimination of zoonotic diseases."* The Department of Health, Department of Agriculture, and the Department of Environment and Natural Resources have formed PhilCZ. PhilCZ has agreed to work on five priority diseases: rabies, anthrax, leptospirosis, Japanese B Encephalitis and Ebola Reston. They also have joint

---

[70] 2018 WHO Joint External Evaluation.
[71] Republic of the Philippines, Office of the President, *Executive Order 10*, 2011.

teams collaborating at central and regional levels, and border points of entry into the country, to manage the risks of outbreaks and spread. However, beyond these areas, PhilCZ has not evolved significant plans and strategies for managing zoonotic disease and does not have a plan for the management of an epidemic or pandemic outbreak arising from a novel pathogen.

> **Improve the functionality and effectiveness of local government units in zoonosis control and prevention, heighten public health literacy, and enhance public health campaigns.**

118.   **The lack of geographical diversification of epidemic and pandemic risk is a central risk management challenge (The Geneva Association 2021).** From a technical point of view, the insurance of pandemic-related losses is problematic as geographic diversification of this risk, the central principle on which the benefits of insurance is based, is not possible. By the sheer definition of epidemics and pandemics, diversification of the risk within a country or, as seen in the case of COVID-19, globally is not possible, making geographic diversification not applicable for the management of this risk. Capital market investors, too, are likely to steer clear of pandemic risk transfer solutions, given their correlation with financial market impacts.

119.   **Time diversification of epidemic and pandemic risk is possible due to the low frequency of severe epidemics and pandemics but the commercialization of those products is challenging.** The diversification in time requires the design of multi-year policies. Such policies themselves present important obstacles in their commercialization. Policies with a duration of more than a year are very seldomly written under the property line of business due to volatility in frequency and severity and the commitment over years to pay for a premium is demanding.

120.   **A complex payout design in line with the insurable interest principle is necessary.** In addition to the multiyear duration of the policies, the payouts on a policy needed for the design of an effective insurance product for protecting the government budget will require a mechanism in the design of the product that matches the financial demands generated by epidemics and pandemics. Thus, probably several payout triggers might be necessary in the presence of a declared epidemic or pandemic. Possible triggers will have to be related to situations requiring financial intervention, such as the level of bankruptcies or unemployment or the bed hospital occupation rate. Designing such triggers that are objective, transparent, and not prone to fraud will need special dedication.

> **Pandemic and epidemic risk transfer instruments which provide funds at the points of need should be acquired by the government when available.** *The sole utilization of development loan facilities, while an important element of the recent response, may not be the optimum way of financing future events as they postpone development projects and may require lengthy negotiations with the development partners. At the same time, the availability of funds that are triggered by the presence of preventive healthcare needs will allow access to therapies in a timely way. Following the risked-layered approach to finance disasters, insurance and insurance linked securities could be designed by leading (re)insurers for this purpose if demand exists (also recommendation under 3.4.8. Diagnostic and Recommended Actions of the Business Interruption Insurance, para. 184).*

3.2.5 Agriculture Risk Protection

Overview of Agriculture in the Philippines

121.   **The contribution of agriculture to the GDP of the Philippines has remained around 10% (10.2% in 2020) since 2016.** This is quite low compared to the share of industry (around 30%) and services (around 60%). Despite its modest share in GDP, agriculture in the Philippines employs a sizable share of the total workforce. In 2020, almost 9.75 million people were employed in agriculture—24.8% of the total workforce. The share of agriculture in employment, however, is slowly declining. In 2015 agriculture employed 11.06 million people, almost 27% of the total workforce (PSA 2021).

122.   **Within agriculture, forestry, and fishing, agricultural crops contributed more than half of the gross value-added (GVA) over the 5 years through 2016.** GVA in agriculture, forestry, and fishing grew between -1% and +1% in the 5 years, except for in 2017, when the sector grew 4.2% (PSA 2021). While agricultural crops were 51.4% of GVA in 2020, fishing and livestock were 12.5% and 12.2% of GVA, respectively (PSA 2021). Among the agriculture, forestry, and fishing subsectors, while agricultural crops grew 1.5% nominally in 2020, livestock and fisheries declined 6.9% and 1.3%, respectively (PSA 2021).

123.   **The agricultural trade balance over the 5 years through 2016 has been negative.** The Philippines exports commodities such as abaca, coconut oil, pineapple products, tuna, etc.; and imports rice, wheat and meslin, milk and cream products, soyabean oil/cake meal, etc. However, agricultural exports have been roughly only half of imports, rendering a negative agricultural trade balance over the 5 years (PSA 2021). By volume of production, the main crops produced in the Philippines in 2020 were sugarcane, palay paddy (unmilled rice), coconut, and banana. By value, the top four crops were palay, banana, corn, and coconut. By planted area, palay, coconut, and corn were the top crops (PSA 2019). The cropped area as well as production of high value crops such as casava, mango, pineapple, coffee, etc., is currently very small but steadily increasing.

124.   **Like many other developing countries, the Philippines also has many smallholder farmers.** Data on agriculture landholdings show that more than 56% of the total number of farms/holdings own less than 1 hectare of land. Thus, more than half of farm owners possess less than 1 hectare of land. However, this large number of small farmers together own only just over 12% of the total agricultural land in the country. This also means that almost 88% of the total agricultural land in the Philippines is owned by 44% of farm owners (PSA 2015).

125.   **The Department of Agrarian Reform encourages selected farmers to own land.** Under its Comprehensive Agrarian Reform Program, it undertakes the transfer of arable land to the ownership of qualified farmer-beneficiaries. Since 1972, almost 4.84 million hectares of agricultural land has been transferred to almost 2.9 million agrarian reform beneficiaries under this program (PSA 2019).

126.   **The Agricultural Credit Policy Council under the Department of Agriculture governs credit policy for the agriculture sector.** Agriculture credit is provided both by private and public sector banks. In the private sector, private commercial banks, savings and mortgage banks, private development banks, rural banks, and stocks, savings and loan associations offer agriculture loans while in the public sector the Development Bank of the

Philippines and the Land Bank of the Philippines are involved. In 2018, private sector banks offered 81.4% of the total agriculture production loans, worth ₱502.6 billion, in the country, while public sector banks provided the remaining 18.6%. However, the share of private sector banks in agriculture has been declining since 2015. The growth rate of total agriculture production loans in 2018 was 14.87% against 7.92% in 2017, compared to 23%, 54.8%, and (-)3.3% in 2014, 2015, and 2016, respectively (PSA 2019).

127.   **The overall rate of credit growth in the country slowed to 1.1% in the third quarter of 2020 compared to the same period in 2019, perhaps due to the impact of COVID-19 pandemic lockdowns.** Correspondingly, the share of agriculture loans in the total loan portfolio also declined, from 6.1% in first three quarters of 2019 to 5.8% during the same period in 2020. However, despite (or maybe because of) the pandemic, within the agriculture portfolio, loans for both agricultural production and transportation and storage grew 6% in 2020 from previous year's level. The increase in loans for production, storage, and transport loans for agricultural products was abetted by both government and private sector credit interventions aimed at securing the supply of food during the pandemic (Agricultural Credit Policy Council 2020).

128.   **Banks are mandated to provide credit to the agriculture sector.** The Agri-Agra Reform Act of 2009 (RA 10000), commonly referred to as Agri-Agra Law, mandates all banks in the Philippines to lend a specified percentage of their loanable funds to the agriculture sector. Accordingly, the banks are required to set aside 25% of their loanable funds for the sector. Within this, at least 10% of total loanable funds must be allocated for agrarian reform beneficiaries and 15% for farmers, fishers, and agriculture in general. Banks can comply with this law by either extending credit directly to eligible classes of beneficiaries or, alternatively, by buying debt securities issued by Land Bank of the Philippines, Development Bank of the Philippines or other debt securities approved by Department of Agriculture or by subscribing to the equity of accredited rural financial institutions or the PCIC.

129.   **However, the banking sector has not been able to fully comply with these statutory requirements ever since the law came into force in 2011.** In fact, the compliance levels have been declining. In 2020, as against the mandatory 25% allocation of loanable funds to Agri-Agra, the banking sector could only allocate 10%, which was lower than the previous year by 2 percentage points (Agricultural Credit Policy Council 2020). Failure to comply with these mandates attracts a penalty of 0.5% of the amount of noncompliance or under-compliance. Ninety percent of the penalties so collected are shared equally between the Philippines Guarantee Corporation (PhilGuarantee) and PCIC. The remaining 10% goes to Bangko Sentral ng Pilipinas (BSP). In 2019, total penalties collected were ₱3 billion.

130.   **PhilGuarantee administers the Agricultural Guarantee Fund Pool.** The pool has been set up to provide credit guarantee on eligible agriculture loans extended by accredited lending institutions. Credit guarantee is provided up to 85% on unsecured loans extended by accredited lending institutions against all types of risks including natural perils of non-repayment by farmer- and fisherfolk-borrowers, including nonpayment due to disasters. Nonpayment due to fraud or willful misrepresentation is excluded from the cover. The objective is to encourage banks and other lending institutions, such as cooperatives, farmer organizations, nongovernment organizations, and corporations, to increase their loans to the agri-agra sector by lowering lenders' risks in a non-collateralized lend. The guarantee fee (premium) varies from 1.00% to 3.85% of the loan amount. In 2020, guarantees

for 60,565 loan accounts with loans amounting to ₱4.25 billion were issued by PhilGuarantee (Agricultural Credit Policy Council 2020). During the pandemic, PhilGuarantee reduced its guarantee fees from 1.0% to 0.5% and increased the guarantee coverage from 85% to 90% for the agriculture sector.

131.   **Disasters significantly affect the agriculture sector.** As noted, the Philippines is struck regularly by meteorological, hydrological, and geophysical disasters. Between 2008 and 2018, 68 tropical cyclones struck the country, affecting a total of 290,812 families (PSA 2020). Total losses to the agriculture sector due to all types of natural hazard from 2010 to 2019 were ₱244.8 billion, equivalent to about ₱24.5 billion annually. During the same period, the total disaster losses were ₱408.9 billion. Thus, the agriculture sector accounted for almost 60% of the reported total losses inflicted by disasters (PSA 2020). From another perspective, average annual losses to agriculture sector works out to 1.37% of total GVA in agriculture, forestry, and fishing for the year 2020 and 30% of total budget allocation of ₱80.4 billion for the agriculture sector in 2020.[72]

132.   **A very small portion of the total planted area in the Philippines is irrigable.** Out of the 13.45 million hectares of total planted area in the country, the estimated total irrigable area is only 3.13 million hectares. Almost 64% of the total irrigable area has been irrigated so far (service area). The total national government spending on agriculture and agrarian reform was ₱147.332 billion in 2020, which was 3.6% of total national government spending. This percentage has remained around 3.5% over the 5 years through 2016.[73]

133.   **The National Food Agency is mandated to procure palay directly from farmers and farmer organizations at government support prices.** This is geared to maintain buffer stocks of rice for emergencies and to sustain government disaster relief during calamities. The National Food Agency is also mandated to distribute rice to the Department of Social Welfare and Development, local government units, and other agencies involved in relief during emergencies and is allowed to dispose its inventory through any authorized mode of disposition before the quality of stocks starts to deteriorate. The government procured 683,132 metric tons of palay in 2020 and distributed 590,696 metric tons of rice (footnote 73).

3.2.6 Diagnostic and Recommendations

134.   **Agriculture insurance is heavily subsidized around the world, including the Philippines.** This is mainly because, traditionally, agriculture has been an unrewarding livelihood featuring high poverty levels. A 2012 poverty study (Reyes et al. 2012) estimated that poverty incidence in the Philippines among agricultural households (57%) is almost three times that of the nonagricultural households (17%).

---

[72] Department of Budget Management. 2020 Budget-at-a-Glance. 2020-Budget-at-a-Glance-Signed.pdf (dbm.gov.ph).
[73] Philippine Statistics Authority. Agriculture Indicators System, Government Support to Agriculture Sector, 2021–2022.

135.   **These well-intended subsidies, however, disincentivize individual risk management.** Given free insurance premiums and the choice of what farming plot to insure, farmers tend to choose the plot with the highest probability of damage. Because of this, farmers tend to use less fertilizer in the plots insured freely (PIDS 2017). Cole, Gine, and Vickery (2014) on financial innovation of small Indian agricultural producers found out that while insurance provision has little effect on total agricultural investments, it significantly shifts the composition of investments toward riskier production activities (PIDS 2017). Insurance should de-risk farming rather than induce the farmers to take up riskier farming activities. Hence while subsidies have helped poor farmers, they have not really induced them to reduce risk and adopt better farm management approaches (PIDS 2017) or adoption of high-value commercial crops, or proper crop rotation. Smart subsidies can be used to make the farmers more receptive toward risk management at the farm level to achieve the overall de-risking of the sector.

To initiate the transformation of agriculture from a paltry livelihood into a sustainable business, huge behavior change must be undertaken. *In particular, the way agriculture insurance, especially subsidized insurance, is viewed by farmers needs a fundamental shift. The presence of insurance should improve farm level risk management. For this, longitudinal farmer education programs must be undertaken and equally important, the right incentives (and disincentives) offered. Subsidies need to be used smartly to usher the much-needed behavior change among farming communities. For example, subsidies can be made contingent upon or proportionate to the verifiable risk reduction measures adopted by individual farmers.*

3.3 Credibility of the Private Sector Offering Risk Transfer Solutions

3.3.1 The Insurance Regulation and Supervision

136.   **Insurance regulation and supervision in the Philippines is moving toward a high observance of the Insurance Core Principles (ICPs).** A 2019 Financial Sector Assessment Program conducted by the World Bank on the insurance sector showed important improvement toward the observance of the International Association of Insurance Supervisors ICPs (World Bank 2021). The Philippines Risk Based Capital (RBC2) standard draws on international practice and experience in the Philippine market to define risk sensitive solvency requirements. The minimum capital requirement is one of the highest in the region and governance and risk management requirements have been strengthening.

137.   **The operational independence of the Insurance Commission suffers from certain limitations.** Supervisory operational independence is required for the intrusive, skeptical, proactive, comprehensive, adaptive, and conclusive supervision necessary for establishing and maintaining a sound insurance sector (Elliott, Erbenova, and Pancorbo 2013). Impediments to act with operational independence have been identified in the ICP 2. Several of these are present in the Insurance Commission:

- The insurance commissioner and deputy insurance commissioners in the Philippines are appointed by and serve at the need of the President.
- The Insurance Commission is a government agency under the Department of Finance and is funded by annual budget allocations. Staff are employed under the same rules as other public servants and at rates of remuneration well below industry counterparts.
- Staff are not provided with the legal protection required by the ICPs (although practice is that staff receiving an action against them relating to their duties are represented by government solicitors).

138.   **Insurance usage remains low.** Despite government efforts, overall insurance penetration in the Philippines continues to stagnate around 3% of GDP since 2015 (at constant prices). The trend is similar for insurance density. Total premiums as a percentage of gross national income have also remained around 1.3% in past few years.[74] This indicates slower growth of the insurance industry compared to the overall economy, as in many other developing economies. On the other hand, insurance policy and regulatory supports to the microinsurance sector have increased microinsurance penetration among the low-income people.

139.   **Among the regulatory challenges that the industry is facing is the next increment of minimum capital and implementation of International Financial Reporting Standards (IFRS) 17:**

- Under the Republic Act 10607, or the Insurance Code, new established insurers are required to have ₱1 billion in paid-up capital, while existing insurers must have a minimum capital of ₱900 million by 31 December 2019 (only very few insurers have complied with this requirement as of the end of 2021) and ₱1.3 billion by 31 December 2022.
- IFRS 17 was originally set for implementation in 2021, but was first deferred to 2023 to allow insurers to revise their timetable and remove any risks from the massive implementation of the project. The Insurance Commission issued Circular Letter 2020–62 further deferring the IFRS17 implementation by 2 years after its effective date given COVID-19 stresses on the sector and economy. This is construed to mean that the local implementation date of the new standards will be 2025.

3.3.2 Diagnostic and Recommended Actions

140.   **Operational independence as stated under the International Association of Insurance Supervisors ICP 2 is central for effective supervision, but the existing legislation and structure does not observe this principle.**

> **Improving the operational independence of the Insurance Commission should be accompanied by measures to increase its formal accountability to the government.** *It is recommended that the authorities review the legislative framework to strengthen the Insurance Commission's formal and financial independence, which should be accompanied by measures to increase its formal accountability to the government.*

---

[74] Insurance Commission. Key Statistical Data 2015–2019.

141.   **The existing capital regime RBC2 captures the insolvency risks, lowering the relevance of the minimum capital requirement.** The upgraded RBC2 regime well reflects the risk that insurers take providing a prudential level of funds to withstand adverse developments in the portfolio. Therefore, the minimum capital requirement becomes only important to guarantee that new insurers or those with low business volume have sufficient funds to operate insurance in a thoughtful manner.

> **Maintaining the minimum capital requirements at the current level of ₱900 million but strengthening the monitoring and compliance with the RBC2 requirements is recommended.** *Under the modern European Solvency 2 regime, the required minimum capital for nonlife companies is €2.5 (₱150 million) and €3.7 million (₱216 million) for life and reinsurance companies, providing indication that the minimum capital does not need to be excessive to have a sound solvency framework, thus supporting to consider revising the scheduled minimum capital increment to ₱1.3 billion by the end of 2022. Further increasing minimum capital requirements may only reduce competition in the insurance sector, leaving less insurance options to the public without having much impact on sector resilience if RBC2 monitoring and compliance is strengthened.*

3.3.3 The Insurance Sector

142.   **The insurance sector seems to have weathered COVID-19 pandemic well.** As indicated in Table 10, premiums grew 20% in 2021, compared with the 2020 performance as reported by the Insurance Commission, outperforming GDP growth. This saw insurance penetration jump to 1.93% by the end of 2021 from the 1.71% in 2020. The life premium remains 2.5 times higher than the nonlife premium. Profitability also increased, by almost 20%, notwithstanding the 40% increase in benefits and claims payments. The capital or net worth of the sector grew 8%, amounting a total of ₱396 billion. Three new companies began operating in 2021, reaching a total of 136 entities operating in the market.

Table 10: Insurance Sector Indicators
(₱ billion)

	2018	2019	2020	2021
Total assets	1,579	1,785	1,929	2,091
Total liabilities	1,241	1,403	1,563	1,695
Capital/net worth	338	382	367	396
Total investments	1,329	1,594	1,705	1,839
Total premium	290	305	308	375
Total benefits and claims paid	103	112	99	140
Total net income	38	45	41	48
Insurance density (₱)	2,722	2,813	2,825	3,400
Insurance penetration (% of GDP)	1.67%	1.64%	1.71%	21.93%

GDP = gross domestic product.
Source: Insurance Commission Statistics.

143.   **The insurance sector response to COVID-19 pandemic resulted in an increased awareness of insurance benefits.** Pandemic exclusions were waived, premium holidays provided, and *ex-gratia* COVID-19 related payments allowed. At the same time, remote underwriting and marketing was developed, providing offers to the public that were at large in lockdowns or working from home. COVID-19 related claims reached ₱3.89 billion as of the end of 2020. The HMO sector accounted for 49% of the said amount, life insurers for claims amounting to 38%, and the rest covered by mutual benefit associations (MBAs) (9%) and nonlife insurers (4%). Members of PIRA also unanimously expressed their belief that the COVID-19 crisis is ushering in a world of new opportunities for the insurance industry for new products and new ways of doing business with IT upgrades and developments.

144.   **Three state-owned and controlled insurance entities are critical players in providing insurance to the country:**

- The Government Service Insurance System (GSIS) provides insurance to several government programs.[75] GSIS is a social insurance institution that provides a defined benefit scheme under the law. It insures its members against the occurrence of certain contingencies in exchange for their monthly premium contributions. GSIS members are entitled to an array of social security benefits, such as life insurance, separation or retirement, and disability benefits. GSIS is also the administrator of the General Insurance Fund (also known as Property Insurance Fund) by virtue of RA 656 (the Property Insurance Law). It provides insurance coverage to government assets and properties that have government insurable interests. The principal benefit package of the GSIS consists of compulsory and optional life insurance, retirement, separation, and employee's compensation benefits. In addition, GSIS offers: multipurpose loans, policy loans, enhanced emergency loans, enhanced pension loans, and pensioners emergency loans.
- PCIC is the implementing agency of the government's agricultural insurance program.[76] PCIC is a government owned and controlled corporation (created by Presidential Decree 1467 (11 June 1978) attached to the Department of Finance, by Executive Order 148 (14 September 2021). The PCIC's principal mandate is to provide insurance protection to farmers against losses arising from natural calamities, plant diseases, and pest infestations of their palay and corn crops, and other crops. PCIC's programs for noncommercial farmers are subsidized. PCIC's initial capital was sourced through the Agriculture Guarantee Fund then managed by the Land Bank of the Philippines (section 3.3.5 for PCIC details).
- The Philippine Health Insurance Corporation (PhilHealth) is providing the universal health insurance program.[77] PhilHealth is a tax-exempt government corporation attached to the Department of Health for policy coordination and guidance. It was created in 1995 to provide the National Health Insurance Program (section 3.2.3 para. 99 onwards for more details on PhilHealth).

---

[75] The Government Service Insurance System at https://www.gsis.gov.ph.
[76] Philippine Crop Insurance Corporation at https://pcic.gov.ph/about-us/ for more information.
[77] Philippine Health Insurance Corporation at https://www.philhealth.gov.ph for more information.

145.   **The financial condition of the three entities is poor and deteriorating.**

- GSIS: Assets amounted to ₱1,436 billion and ₱2,020 billion liabilities resulting in negative equity of ₱606 billion, increasing from a deficit of ₱500 billion in 2019. Comprehensive income was negative ₱108 billion, a significant deterioration of ₱246 billion over 2019's ₱138 billion profit.
- PCIC: The 2020 financial statement shows assets totaling ₱6.398 billion, down 15% from 2019, and liabilities amounting to ₱3.348 billion, down 26% from 2019, resulting in capital of ₱3.049 billion, almost the same as in 2019. Income was ₱634 million, down 34% from 2019. Note that the 2020 audited financial statement was qualified.
- PhilHealth: The latest published financial statement is for 2019. It indicates assets of ₱221 billion, a 25% increment from 2018; and ₱111 billion in liabilities, an increment of over 50%; resulting in ₱110 billion net worth, slightly above 2018 net worth. The reported income in 2019 was ₱4.3 billion, down from ₱21 billion in 2018.

146.   **In August 2020, the Department of Justice was directed to investigate governance concerns at PhilHealth. A taskforce has been set up to conduct the investigation, which is ongoing.** The taskforce, headed by the Department of Justice, is composed of the National Bureau of Investigation, Presidential Anti-Corruption Commission, the Office of the Special Assistant to the President, National Prosecution Service, and the Anti-Money laundering Council. The taskforce has started investigations into at least 27 cases of fraud and corruption, but interim findings and recommendations have not yet been published. Nonetheless, the scope and seriousness of the investigation has impacted the credibility of the state insurer, and perhaps highlighted issues in the largely self-regulatory approach to holding PhilHealth to account. PhilHealth's Board, through its constitution, is chaired by the secretary for health and, through this mechanism, PhilHealth is accountable to the government. However, recognizing that PhilHealth is a financial entity dealing with fund management, Senate Bill No. 1829 was filed in 2020,[78] which seeks to amend the Universal Health Care Act and allow the appointment of the Department of Finance secretary as chair of the PhilHealth board. The bill has yet to be passed (March 2022).

3.3.4 Diagnostic and Recommended Actions

147.   **Key participants in the provision of insurance to the public— Philippine Crop Insurance Corporation, Government Service Insurance System, and PhilHealth—lag behind compliance with insurance technical practices and need effective insurance supervision:**

- GSIS: GSIS is an institution providing social insurance, responsible for certain benefits for public employees and insuring public assets that are not regulated by the Insurance Commission. GSIS is in direct competition with the private sector, for instance in the insurance of public assets, and thus the different regulation and supervision enjoyed by GSIS results in an unlevel playing field. This situation could motivate Insurance Commission regulated competitors to take excessive risk trying to compete with GSIS. Also, the negative equity of the GSIS needs close attention to maintain the viability and soundness of the institution.

---

[78] Senate of the Republic of the Philippines. *Senate Bill 1829: Amending section 13 of the Universal Health Care Act*, 2020.

- **PCIC:** The Insurance Commission analyzed the financial status of the PCIC in-depth in 2021, showing significant exposure to major catastrophes. Risk management and internal controls were seen as deficient and the reporting of financial statements found not to provide a clear picture complying with the Philippine Financial Reporting Standards for insurance companies. In a major governance reform, PCIC has been placed under the supervision of Department of Finance, instead of the Department of Agriculture, through an Executive Order of 14 September 2021. As a result of this change, the PCIC Board will now be chaired by the Secretary Department of Finance with the general manager of GSIS as another board member. In a report to the Department of Finance, the Insurance Commission recommended that the PCIC revisit its overall risk management program and include reinsurance among its risk transfer mechanisms to better manage its risks and avoid the scenario of heavy capital infusions from the government to cover possibly large-scale losses in the future.[79]

- **PhilHealth is self-regulated:** PhilHealth and the Insurance Commission have agreed to a cooperative arrangement, with PhilHealth voluntarily agreeing to the "PhilHealth examination," a 6-month comprehensive assessment of key insurance processes. This was agreed in 2019, so it is perhaps too early to determine its effectiveness. However, audit reports have—in recent years—highlighted a number of operational challenges with PhilHealth, including potential fraud, waste, and abuse in excess of what might normally be expected with adequate controls.

Effective insurance supervision is critical for the financial resilience of institutions undertaking insurance activities. *Given that the findings on GSIS, PCIC, and PhilHealth show they are lagging behind compliance with insurance technical principles that could be exacerbated in major disasters and/or another pandemic, its supervision of their insurance activities should be assigned to the Insurance Commission.*

3.3.5 Agriculture Insurance

148.   **Agriculture insurance is provided almost exclusively by the PCIC.** The PCIC has been mandated among other functions to provide insurance protection to agricultural producers in the Philippines against losses of crops and non-crop agricultural assets due to natural calamities, pests and diseases, and other perils. It implements and manages various government agricultural insurance programs. Under the auspices of the Department of Agriculture, the PCIC operates as a government-owned and controlled corporation and its administrative operations are not funded by the national government (PIDS 2014). PCIC's operating expenses are funded from its own corporate operating budget. In the Philippines, however, agricultural insurance has been viewed as something more than just a risk management tool. While the aforementioned concept of agricultural insurance is adopted in RA 8175, both PD 1467 and PD 1733 referred to the said tool as crop insurance and stated that it can serve as "a relief good" to crop producers whose farms are adversely affected by natural calamities or other perils. Thus, the concept of social insurance was also introduced.

---

[79] For more information, see the Department of Finance, New and Views for 01 October 2021 at https://www.dof.gov.ph.

149.   **A large portion of agriculture insurance offered by PCIC is fully or partly subsidized through government premium subsidies.** PCIC offers a full premium subsidy to beneficiaries in the Registry System for Basic Sectors in Agriculture (RSBSA) for rice and corn crops, while partial subsides are offered to other farmers under various programs of the Department of Agriculture. Box 4 looks at the RSBSA. Subsidies are financed through the General Appropriation Act, Department of Agriculture's budget allocations as well as PCIC's own funds raised from agri-agra penalties. Note that PCIC also offers nonsubsidized programs aimed at commercial farmers. Penetration is limited.

Box 4: Registry System for Basic Sectors in Agriculture History

The Registry System for Basic Sectors in Agriculture (RSBSA) project, initiated in 2012, aims to identify farmers and fisherfolk that need to benefit from agriculture-related programs and services of the government. This registry is used as the basis for the Department of Budget and Management to target beneficiaries of some of the different agricultural support programs of the government as implemented by various government agencies, such as the RSBSA agricultural insurance program of the Philippine Crop Insurance Corporation (PCIC), and the Agriculture and Fisheries Financing Program of the Land Bank of the Philippines and People's Credit and Finance Corporation.

For the RSBSA, the basic sectors in agriculture refer only to crop and animal production, aquaculture, and fishing. Apart from farmers, it also includes farm laborers. Activities related to hunting, forestry, and logging are not considered basic sectors in the registry. Registration is carried out through house-to-house surveys and validation of the applicant's eligibility is done by the respective line agencies. Since the RSBSA contains self-reported data, many inconsistencies are likely. The RSBSA list does not tally with the list of beneficiaries under various programs maintained by agencies like the Department of Agrarian Reform and PCIC.

There is a feeling that the RSBSA may have left out many eligible farmers and fisherfolk. Efforts have been made to crossmatch databases from different agencies with RSBSA, but standardized reference database and uniform protocols for inclusion and exclusion are absent.

Despite its shortcomings, the RSBSA is quite a useful tool as a starting point for the creation of a unified government database in the agriculture and fisheries sector.

The RSBSA enlists about 10.91 million beneficiaries, out of which 5.48 million are farmers, 1.36 million fisherfolk, and 4.06 million farm laborers.[a]

[a] PCIC. Revised Implementing Guidelines on the Utilization Of Government Premium Subsidy to the Philippine Crop Insurance Corporation Under FY 2017 General Appropriations Act, RA 10924. https://pcic.gov.ph/rsbsa/.

Source: Authors.

150.   **The subsidy bill more than doubled from ₱2.5 billion in 2017 to ₱5.5 billion in 2019, before declining to ₱4.8 billion in 2020 and ₱4.5 billion in 2021.** The decline in government premium subsidies (GPS) in 2020 appears to be in line with the government policy to phase out subsidies to focus on easy access to credit for agriculture production (Simeon 2018). A bill filed in the Philippines Senate in 2019 seeks to offer a (fully subsidized) index-based insurance to poor rice farmers (Senate of the Philippines 2019). The bill also seeks to provide a premium subsidy of ₱6 billion to cover over 2.8 million hectares of rice land. In 2021,

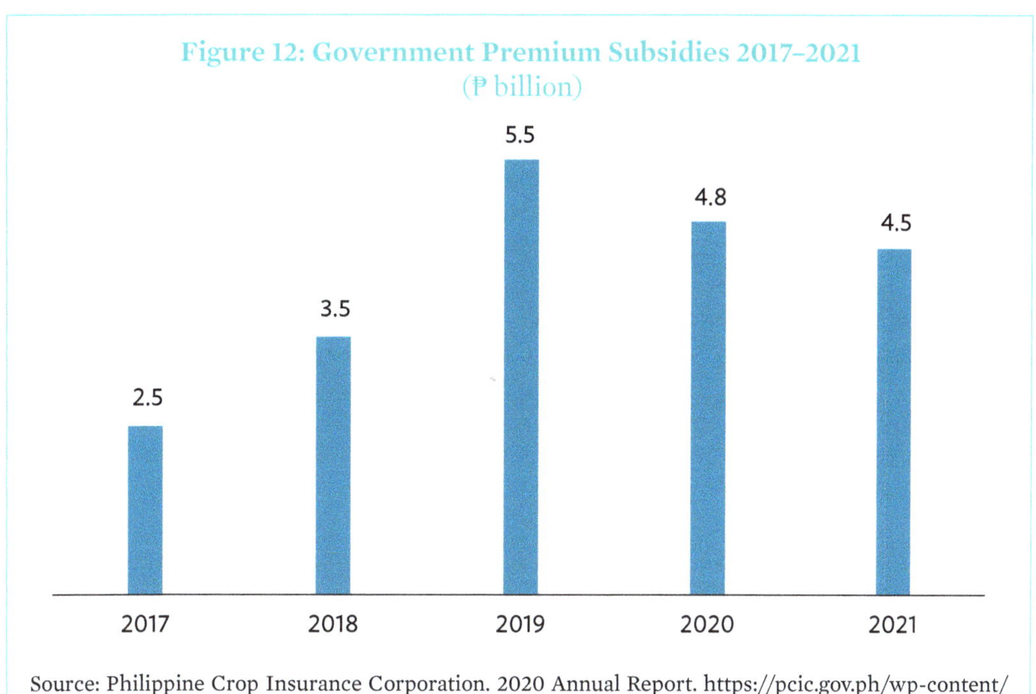

Source: Philippine Crop Insurance Corporation. 2020 Annual Report. https://pcic.gov.ph/wp-content/uploads/2021/10/annual-2020-for-website-final.pdf.

the Agriculture and Livestock Crop Insurance Program was allocated ₱4.5 billion to give full insurance premiums to 2.1 million subsistence farmers and protect them against losses of crops and non-crop agricultural assets (Figure 12).

151.   **A further ₱7.0 billion allocation was made available for the buffer stock program of rice supplies for emergencies.** Under the Social Protection Program there are specific allocations of funds for "swift" response to crises.[80]

152.   **A large part of PCIC's premium portfolio comprises subsidized insurance offered to RSBSA beneficiaries and other target groups.** In 2020, out of the over ₱5.086 billion in premiums generated, about 94.84%, or ₱4.824 billion, consisted of GPS, while the balance of about 5.16% of total premium consisted of premiums paid by farmers, fisherfolk, and lending institutions, worth a little over ₱262 million (PCIC 2020). Over 3 million farmers were covered under PCIC's various insurance programs in 2020, out of which 41.38% were rice farmers and 13.78% corn farmers; 21.97% of farmers were covered under term life insurance. Livestock and high-value commercial crops farmers comprised 11.98% and 9.22%, respectively. In premium distribution, 60% of PCIC's total premium income was toward rice insurance (PCIC 2020). In 2020, PCIC paid claims worth ₱3.5 billion to over 611,000 farmers. The overall loss ratio of PCIC in 2020 was 69% (PCIC 2020).

---

[80] See page of the 2021 People's Enacted Budget (dbm.gov.ph).

153.   **PCIC insurance protection to fish farmers/fisherfolk/growers is small.** The cover protects against losses in unharvested crop or stock in fisheries farms due to disasters and fortuitous events. The insurance covers the cost of production inputs, the value of the fish farmer/fisherfolk/grower's own labor and those of the members of his/her own household, including the value of labor of hired workers per the Fisheries Farm Plan and budget. In 2020, 47,332 farmers were covered under Fisheries Insurance, with a premium income including subsidies of ₱76.8 million, roughly 1.5% of PCIC's total premium.

154.   **Livestock insurance includes swine, which topped the list at 456,226 heads in 2020.** Poultry followed with 338,025 heads. Cattle and goat had nearly 180,000 head each. In 2020, livestock insurance covered 1.3 million heads. There were over 11,000 horses and about 1,700 sheep. The total premium collected inclusive of GPS and farmers' contribution for livestock insurance in 2020 was ₱378.2 million, which amounts to 7.4% of PCIC's total premium.

155.   **Notwithstanding the involvement of PCIC, available data on economic losses to the agriculture sector reveals a huge protection gap.** Tropical cyclones, floods, and droughts have been the top-three risks, to which Philippine agriculture is exposed (PIDS 2017). Based on data available for 2003–2013, average annual cost of risk for the agriculture sector from these three hazards was ₱12.9 billion, whereas average annual claim payments by PCIC for the same period were a meager ₱162.8 million, resulting in a huge protection gap of more than 95% (Table 11). While penetration rates have increased substantially, and therefore so have claim payouts in recent years, the protection gap remains significant. For example, the total claim payouts by PCIC in 2018 were ₱3.49 billion, against average annual losses of ₱12.9 billion (2003), leaving a protection gap of about 73%.

156.   **More recent data shows that the protection gap remains large.** Out of the 5.48 million farmers registered under the RSBSA, the PCIC could insure only about 3 million in 2020. In 2020 again, only about 16% of the total planted area could be insured. The coverage for livestock and fisheries has been even more insignificant. Among economic losses suffered by agriculture sector because of disasters, just around 5% of the total losses could be indemnified through claim payouts. Thus, a lot needs to be done to improve outreach and efficacy of agriculture insurance, especially given the country's frequent disasters. If all farmers were to be covered under subsidized insurance schemes, the country would need a subsidy of around ₱30 billion annually. In planted area, the gap is even larger. From 2014 to 2018, the gap between total planted area and total insured area has consistently remained around 93%, with the only exception in 2017 when it narrowed down to 90% (Table 12).

157.   **Notwithstanding the intention to encourage private sector involvement in agriculture insurance, it remains very limited.** The Insurance Commission on 15 October 2021 issued the Guidelines on Adoption of a Regulatory Sandbox Framework for Piloting Agriculture Insurance, which effectively opened the gates for private sector entry into agriculture insurance. The sandbox invites proposals from private sector nonlife insurance companies to submit proposals for agriculture insurance pilots. The proposed pilots can be in collaboration with PCIC or any other reputed national or international organization. Private sector insurance companies can therefore offer agriculture indemnity or index-based insurance products to farmers on a pilot basis. ADB is also simultaneously undertaking a technical assistance project supporting public–private partnership for

agriculture insurance in close consultation with PIRA and NatRe. Under this initiative, CARD Pioneer Microinsurance Incorporated signed a memorandum with PCIC in February 2022 to pilot agricultural insurance for high value crops under a co-insurance agreement. This is part of ADB's TA on financial inclusion and Public–Private Partnership on Agricultural Insurance.

Table 11: Protection Gap in Agriculture, Total Economic Loss vs. Claims Paid

Year	Total Economic Losses to Agriculture Sector[a] (₱, million)	Claims Paid by PCIC[b] (₱, million)	Coverage (%)	Protection Gap (%)
2003	2,998.7	50	1.67%	98.33%
2004	8,803.6	69	0.78%	99.22%
2005	2,272.9	70	3.08%	96.92%
2006	11,607.8	60	0.52%	99.48%
2007	1,696.3	57	3.36%	96.64%
2008	1,4659.1	65	0.44%	99.56%
2009	12,842.7	200	1.56%	98.44%
2010	23,994.7	235	0.98%	99.02%
2011	16,679.4	370	2.22%	97.78%
2012	34,302.8	215	0.63%	99.37%
2013	12,019.5	400	3.33%	96.67%
Total	**141,877.5**	**1791**	**1.26%**	**98.74%**

[a] Losses due to tropical cyclone—destructive, flooding and drought payments.
[b] Figures have been rounded off.
Source: Philippines Statistics Authority. 2020. Compendium of Philippine Environment Statistics, Review of Design and Implementation of the Agricultural Insurance Programs, and PIDS 2014.

Table 12: Protection Gap in Agriculture, Total Planted Versus Insured Area

Year	Planted Area (ha, million)	Area Insured by PCIC (ha, million)[a]	Penetration (%)	Gap (%)
2017	13.5	2.4	17.78%	82.22%
2018	13.5	3.3	24.44%	75.56%
2019	13.3	2.4	18.05%	81.95%
2020	13.4	2.2	16.42%	83.58%

ha = hectares.
[a] Excluding livestock, fisheries, Credit and Life term, insurance, and Non Crop Insurance.
Source: Philippines Statistics Authority. 2020. Selected Statistics on Agriculture 2020. PCIC annual reports.

3.3.6 Diagnostic and Recommended Actions

158.   **Private sector involvement in agriculture insurance remains very limited despite efforts to encourage it.**

Continue encouraging private sector involvement in agriculture insurance, especially for commercial farmers and nontraditional crops.

159.   **Despite PCIC's good work in covering poor farmers, a huge protection gap still exists.**

In view of the above, promoting public–private partnerships for agriculture insurance appears to be a viable option for protection to Filipino farmers in a sustainable manner. *Some beginnings have already been made but much more needs to be done.*

- *A clear policy on public–private partnerships for agriculture insurance needs to be promulgated at the federal level, to convey a clear policy direction.*
- *The private sector needs to be incentivized to enter the agriculture insurance sector, through special regulatory dispensations.*
- *Other incentives could include value-added tax (VAT) exemptions on agriculture insurance premium, tax holidays on income generated through agriculture insurance, attracting private sector investment in infrastructure supporting agriculture insurance.*
- *Revamping RSBSA to better target GPS to eligible farmers.*

3.3.7 Private Health Insurance Sector

160.   **The uptake of private health insurance is rising.** Around 4% of people are covered by an HMO or have fully private health insurance, allowing access to a broader range of providers. Health insurance now accounts for roughly 5% of all life insurance premium income (Figure 13). Currently, 30 HMOs operate with varying levels of maturity, up significantly from 2016, when just 14 HMOs were licensed under the Insurance Commission. Alongside this, 31 life and 58 nonlife insurers are licensed to provide health products. Due to the strong presence of HMOs in the country, the Philippines has not been a prioritized market for most international health insurers, though this may change as affluence grows and, with it, demand for products that can be used internationally. Currently, however, the largest providers of private health insurance are therefore local domestic players (Figure 14, Table 13).

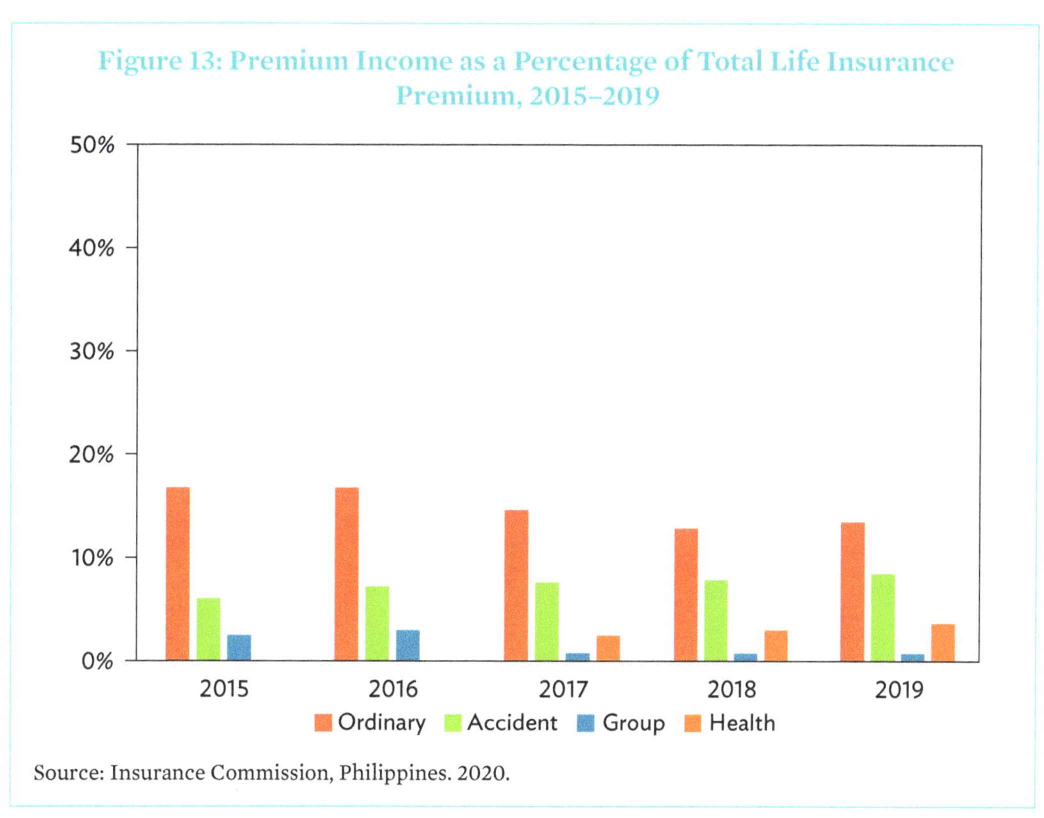

Figure 13: Premium Income as a Percentage of Total Life Insurance Premium, 2015–2019

Source: Insurance Commission, Philippines. 2020.

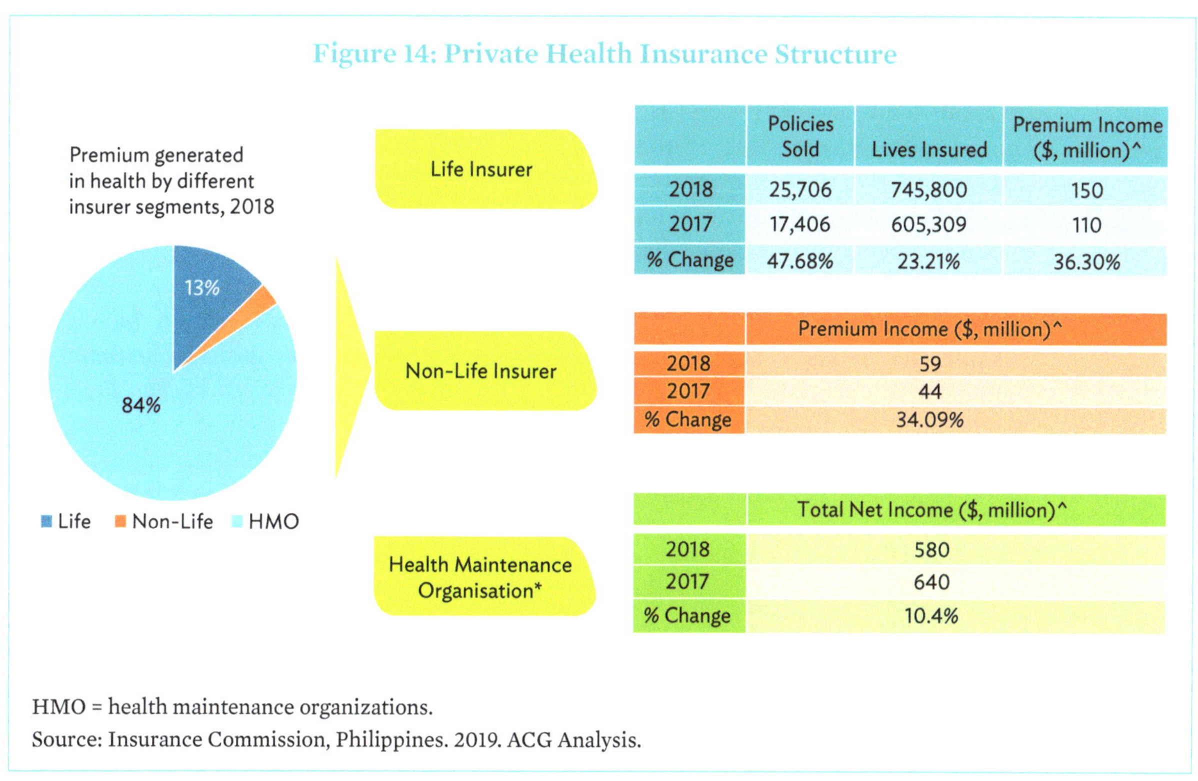

Figure 14: Private Health Insurance Structure

	Policies Sold	Lives Insured	Premium Income ($, million)^
2018	25,706	745,800	150
2017	17,406	605,309	110
% Change	47.68%	23.21%	36.30%

	Premium Income ($, million)^
2018	59
2017	44
% Change	34.09%

	Total Net Income ($, million)^
2018	580
2017	640
% Change	10.4%

HMO = health maintenance organizations.
Source: Insurance Commission, Philippines. 2019. ACG Analysis.

Table 13: Insurers Ranked by Gross Premiums Income in Health in the Philippines, 2019

Rank	Insurers ranked by premium income	Gross premium income - health ($ '000)	Market share (%)	Company/subsidiary company included
1	United Coconut Planters Life*	69,923	41.65	UCPB General Insurance Co. Inc, Cocolife Asset Management Company, Inc.
2	Asian Life*	49,832	29.68	Maybank ATR KimEng Financial Corporation
3	Pacific Cross Insurance (Blue Cross)	15,422	9.19	Pacific Cross Vietnam International Administrators, Ltd., International Services Pacific Cross, Pacific Cross Health Insurance Pcl
4	FWD Life Insurance*	6,618	3.94	FWD Life Insurance Corporation
5	Philippine AXA Life*	5,692	3.39	AXA Group, GT Capital Holdings, Metropolitan Bank and Trust Company (Metrobank)
6	Philippine British Assurance	5,239	3.12	Philippine British Assurance Company (PBAC)
7	Insular Life*	4,018	2.39	Insular Life Healthcare, Insular Life Foundation
8	Sunlife Grepa*	2,402	1.43	Sun Life
9	Fortune Life	1,753	1.04	ALC Group of Companies
10	QBE Seaboard	1,704	1.01	QBE Insurance Group

* Life insurers
₱1 = $0.020
Source: Insurance Commission, Philippines. 2021. ACG Analysis.

161.   **Since 2015, HMOs have been regulated by the Insurance Commission.** Existing HMOs are required to have a minimum paid-up capital of ₱10 million, which increases to ₱100 million for new applicants. Executive order 195, issued in 2015, set out the rules on paid up capital and further required HMOs to provide a deposit to the Insurance Commission equivalent to at least 20% of their paid-up capital in cash, Treasury bills, or bonds.[81] The deposit is intended to protect the interests of HMO members, and ensure sufficient solvency to meet claims costs. Foreign HMOs planning to set up a branch in the Philippines are required to make a statutory deposit of ₱100 million ($2.1 million) in cash or securities.

162.   **Regulatory oversight of HMOs by the Insurance Commission has led to substantial changes in the health insurance industry, with higher levels of financial risk management exercised by HMOs.** Prior to 2015, most HMOs were led by, and organized as, physician groups that offered a form of subscription services to members. There was no requirement for HMOs to utilize epidemiology, service utilization analysis, morbidity and mortality modeling or other actuarial analysis. While balance sheets did have to be submitted by the then de facto regulator—the Department of Health—there was no capacity or capability within the Department of Health to read these accounts, and ascertain financial viability. The Department of Health therefore voluntarily opted to surrender their responsibility for regulating HMOs to the Insurance Commission in 2015; a position which was then brought into law.

---

[81] Office of the President of the Philippines, *Executive order 192 Transferring the Regulation and Supervision Over Health Maintenance Organizations from the Department of Health to the Insurance Commission, Directing the Implementation thereof and for Other Purposes.* 2015 Executive Order No. 192, s. 2015 | Official Gazette of the Republic of the Philippines.

163.   **Since the beginning of the pandemic, the Insurance Commission has moved three HMOs into conservatorship.** These moves reflect efforts to safeguard the interests of members and ensure claim liabilities can be met in an appropriate way. However, as HMOs typically excluded coverage for pandemics and epidemics, and the costs of COVID-19 claims were largely met by PhilHealth and out of pocket spending, the move of HMOs into conservatorship perhaps indicates some financial fragility in the system generally. For health insurers to be effective and viable in the long-term, they typically require a large risk pool. This is because incidences of catastrophic illness, such as cancer or renal failure, can generate large losses that can rarely be recovered over the term of a policy. The large number of HMOs in the Philippines—while offering consumer choice—likely mean risk-pools are not yet sufficiently large or diverse enough to manage claims risks.

164.   **The Insurance Commission has also taken action against aggressive or anti-competitive selling in the HMO sector, particularly during the recent pandemic.** Under Circular Letter (CL) 2021–68,[82] the Insurance Commission instructed HMOs to stop offering any type of discounts on policies sold to individuals and families, such as bulk purchases and promotional markdowns for policies. The Insurance Commission further reminded HMOs that inducements, including the giving of rebates, sharing of commissions or providing any form of valuable gift, were not acceptable practices. The Insurance Commission also highlighted the need for HMOs to exercise non-discriminatory approaches, ensuring Filipinos rights and benefits were on par with those offered to expatriate individuals under the same product-type. This highlights the Insurance Commission's efforts to protect individuals, particularly during periods of vulnerability, and tackle potential corruption and irregularities within the sector.

3.3.8 Diagnostic and Recommended Actions

165.   **The Insurance Commission has been taking adequate actions to increase the credibility in the health sector.** Among other actions, the Insurance Commission has started regulating the HMOs in 2015 and taking disciplinary actions against HMOs, especially during the pandemic, as well as carrying out a voluntarily "PhilHealth examination," highlighting several issues to be resolved.

Insurance Commission should continue efforts to enhance health sector credibility.

3.4 Product Availability and Affordability

166.   **The insurance sector offers a range of products to cover disasters, albeit not covering the whole country**. The insurance sector covers earthquakes, tsunamis, floods, volcano eruptions, and typhoons, among others. However, it abides to strict underwriting guidelines that exclude highly exposed areas where the level of risk would result in unaffordable premiums. Underwriting guidelines also exclude risks that cannot be diversified either in their portfolios or through reinsurance like pandemics, war, nuclear event, etc.

---

[82] Insurance Commission, Circular Letter on "Guidelines on Offering of Discounts on Membership Fees of HMO Products." CL2021_68.pdf (insurance.gov.ph).

167.   **There is no mandatory property insurance.** Private properties are not required to have insurance, except if financed by a bank. Commercial properties are only required to pass an inspection.

3.4.1 Catastrophe Risk Insurance Products

168.   **The significant exposure to disaster losses in the Philippines presents an important challenge for the insurance sector, requiring pooling of resources.** Currently, the insurance sector provides property insurance that covers natural hazards; however, the sector avoids areas where risk is too high for financial capacity, including the available reinsurance programs. The scale of risk exposure can be illustrated by the size of the economic damage ($12.5 billion) caused by Typhoon Yolanda in 2013 which was over 1.5 times the total capital of the whole insurance sector in 2021. As an important step to create disaster risk insurance capacity, the Insurance Commission, Nat Re, and PIRA in January 2020 signed a memorandum of understanding pledging to collaborate toward the implementation of a Philippine Catastrophe Insurance Facility (PCIF). The facility would accept catastrophe risks underwritten by the insurance sector and thus creating national capacity to retain disaster risk and improve the negotiation ability with international reinsures. In addition, it would dedicate resources for disaster risk modeling and pricing. Box 5 provides details the facility.

Box 5: General Framework of the Philippine Catastrophe Insurance Facility

The facility would operate under the following framework:

(i)   implement a compulsory cession of catastrophe risks to the Philippine Catastrophe Insurance Facility (PCIF);
(ii) cessions to PCIF shall be ceded back to subscribing domestic insurance or reinsurance companies based on the proportion of their subscribed amounts and their financial strength;
(iii) the PCIF shall then protect the aggregate catastrophe exposures under a common account excess of loss reinsurance program and do other activities to reduce aggregate reinsurance costs and optimize premium retention and portfolio diversification.

A Technical Working Group chaired by the insurance deputy commissioner would review rates and structure of catastrophe insurance premiums to enable inclusive access to catastrophe insurance protection at technically sustainable rates.

An oversight committee would provide governance function for the PCIF composed of the stakeholders responsible for its vision and sustainability. It would be responsible for setting operating policies, pricing sustainability, product design, underwriting policy, reinsurance cover, and claims standardization. The Insurance Commission, meanwhile, would supply the next level of governance.

Capitalization of ₱1 trillion was envisioned, as of 2021.

Source: 2020–2021 PIRA Factbook; CDRI-Philippine-Catastrophe-Insurance-Facility-Factsheet-2020.pdf.

3.4.2 Diagnostic and Recommended Actions

169.   **The magnitude of disaster losses limits insurance availability in all regions of the country.** A high minimum capital requirement will not increase the disaster insurance availability since the required disaster risk capital would be much higher thus, not even large companies limit their disaster exposure. The establishment of PCIF, if it proceeds, is expected to create within the insurance sector the ability to provide in a technical manner affordable accessible insurance products that cover natural hazards in the whole country.

> **The PCIF should maintain its intended focus, if established, to provide technically sound catastrophe risk insurance available to any part of the country where residential and other buildings are officially allowed to be built.**

3.4.3 Agriculture Insurance Products

170.   **The crop insurance basket of PCIC consists mainly of multiperil crop insurance.** PCIC insurance products are available to cover rice, corn, high-value commercial crops, livestock, and fisheries. PCIC also offers a wide range of insurance products that go beyond agriculture insurance including non-crop agricultural assets and term life insurance for farmers. The crop insurance products offer indemnity for input costs, though for high-value commercial crops, PCIC is piloting products that offer a higher coverage. In addition, some pilots have been carried out for index-based insurance as well, though no clear indication on their scalability and sustainability is available. In 2012, PCIC, in partnership with German Agency for International Cooperation, also tested an Area Yield Index insurance based on remote sensing technology. Apart from PCIC, there are a couple of agri-microinsurance products in the market but their volumes are insignificant. Although PCIC has tried to introduce innovative agriculture insurance products, lack of private sector participation has by and large left a sizable unexplored space for product innovation and market penetration.

3.4.4 Diagnostic and Recommended Actions

171.   **Agriculture involves a wide range of risks with varying frequency and severity.** This makes it practically impossible for a commercial insurer to underwrite all risks associated with the activity. Insurers generally resort to exclusions and deductibles to overcome this limitation, which ultimately ends-up in diminishing the client value of the product and reduced uptake by farmers. The business activities and scope of agricultural cooperatives in the Philippines cover the agribusiness functions including input supply, production, post-harvest, processing and marketing as well as credit and financing. There are no reports of agriculture cooperatives involved in any kind of risk sharing arrangements as of now.

> **This problem can be addressed by a risk-sharing arrangement between the farmers' cooperatives and commercial insurers.** *Products can be designed such that the idiosyncratic risks are borne by the farmers' collectives while the catastrophic risks are carried by insurers. This way, the farmers get a comprehensive coverage while the risks (and therefore premiums) are sliced and distributed at the back end among insurers and cooperatives. This way, the farmers' collectives have an incentive to adopt risk*

reduction measures at the local level and prevent moral hazard and adverse selection. The Philippines has the benefit of having a couple of cooperative insurers who are best placed to take such an initiative.

172.   **Agriculture is exposed to catastrophic risk and therefore reinsurance becomes a critical success factor for agriculture insurance.** To optimize the underwriting capacity of direct insurers, multi-layered reinsurance options must be made available to them. International reinsurance markets can provide the capacity, but they often tend to be volatile and expensive, especially for new products.

An agriculture insurance pool can fill the gap between domestic and international reinsurance markets and can become an important tool for retaining risk domestically. *Apart from the local commercial insurers and reinsurers, the government, banks, and multilateral agencies can also participate in this pool. The pool can be managed professionally under the aegis of a national reinsurer to ensure the buy-in from all stakeholders. Apart from absorbing risk, the agricultural insurance pool can also undertake ancillary activities like data analytics, managing technology platforms and risk research on behalf of the industry. As the Philippines moves toward development of private agriculture insurance market, an insurance pool could serve well to a precursor.*

173.   **If the outreach of agriculture insurance is to be expanded significantly, the farming community needs to have choice among various types of products.** Currently PCIC's portfolio mainly comprise of multi-peril crop insurance but other types of products like area yield index and weather index may be preferred by farmers as well as downstream value chain players like commodity traders, food processing units, exporters and so on. The current set of PCIC products also restrict the coverage to input cost. Farmers would typically want their profits as well to be insured. In fact, the average sum insured per hectare for rice was just ₱27,391 in 2013, (PCIC 2020) whereas the cost of producing palay in the country averaged ₱42,061 per hectare in 2013,[83] which points to an underinsurance of over 50%, even at the input cost.

Product innovations should include value-added services. *Following the prospect theory (Kahneman and Tversky 1978),[84] if farmers are expected to pay for insurance, they would typically expect much more than mere indemnity against contingent losses. To tackle this behavioral anomaly, agriculture insurance products need to be bundled with a variety of value-added services like free cropping advisories, early warning for disasters, discounted inputs, etc., to overcome dissonance and ensure a decent uptake by farmers. Agriculture insurance is more about agriculture than it is about insurance. Hence insurers need to think much beyond simple indemnity in designing agriculture insurance products. Value-added services accompanying insurance products can be neatly crafted such that they are attractive for farmers as well as result in risk reduction for insurers. This way a win-win situation can be created for all stakeholders. Other product innovations like agriculture credit portfolio insurance, value chain insurance, etc., also need to be tried out.*

---

[83] Philippine Statistics Authority. Costs and Returns of Palay Production, 2015.
[84] https://web.mit.edu/curhan/www/docs/Articles/15341_Readings/Behavioral_Decision_Theory/Kahneman_Tversky_1979_Prospect_theory.pdf.

3.4.5 Health Insurance Products Supplementing the Universal Health Care

The range of health protection products offered in the Philippines is large, ranging from comprehensive international private medical insurance through to low-cost micro health insurance. The majority of private health insurance products sold are HMO plans for corporates and are geared toward larger multinational corporations or affluent domestic companies. However, in recent years, considerable effort has been made to stimulate development of health micro insurance products. A regulatory framework for micro health insurance products has been developed, which outlines common provisions proposed for them (Table 14). Insurers who meet these provisions will be eligible to use the micro health insurance logo in product marketing (Insurance Commission 2018).

174. **Private health insurance products typically work alongside the PhilHealth products.** The majority of health products offered by private insurers, including HMOs, build upon and supplement the base product offered by PhilHealth. The product hierarchy does not allow for individuals to opt out of PhilHealth if they have a private health product. If an individual purchases a private product and they are not enrolled in PhilHealth, the element of any healthcare claim that would normally be covered by PhilHealth would need to be borne by the individual. In this regard, products help close protection gaps by reducing the out-of-pocket payments individuals would pay when their PhilHealth benefits and limits were exhausted.

Table 14: Common Provisions for Micro Health Insurance Products

Common Provisions	Description
Microinsurance products maximum premium	7.5% of the current daily minimum wage rate of a nonagricultural worker in Metro Manila, computed on a daily basis
Maximum benefits	1,000 times the daily minimum wage rate of a nonagricultural worker in Metro Manila
Policy Contract	Simple and easy to understand. May be printed in English or Filipino
Frequency of premium collection	Flexible, cash flow based premium payments (daily, weekly, monthly, quarterly, semi-annual, annual)
Documentary requirements	Simple, easy to understand and minimal
Claims settlement	Ten (10) days after submission of complete documents
Contestability period	Maximum of 1 year from date of issue or last reinstatement of the policy
Target market	Low income and informal sector
Providers	Life insurance Companies, nonlife insurance companies, cooperative insurance societies, mutual benefit associations, health maintenance organizations

Source: Insurance Commission 2018.

175.   **PhilHealth clarified that it would be the "first payer" for claims related to COVID-19, paying the full costs of cases that occurred before April 2020, and up to the case-based rate thereafter.** There was initial confusion over both the coverage of COVID-19 by insurance products and the order in which individuals and providers should or could avail insurance benefits. PhilHealth confirmed in early 2020 that it would be the first payor, and cover the total costs of claims incurred for COVID-19 on a fee-for-service basis. However, in April 2020, PhilHealth introduced case-based payments to providers, which paid a set fee for COVID-19 cases. This meant any residual balance had to be covered by individuals, either through any supplemental HMO or private insurance policy held or out of pocket.

176.   **Commercial insurers sought to cover the costs of the recent pandemic despite epidemics and pandemics universally being excluded by policy terms.** This was largely seen as a goodwill effort to support members and help retain health insurance policies during a period of economic uncertainty. The extended benefits offered were often provided at either no additional cost to members or through a small optional rider. Benefits offered were typically capped to shield insurers from the most catastrophic of costs, such as long-term confinement costs for patients in intensive care.

177.   **Many insurers offering critical illness plans also updated these to include both COVID-19 and a range of other communicable diseases, including Ebola and Zika.** This shows that with good global epidemiology and cost data, insurers can work with some speed to incorporate a novel disease into products and coverage. There is often a commercial benefit in doing so; pandemics and epidemics generally increase sales of private protection products and competition means insurers will often provide some cover to attract and retain policyholders.

178.   **Future novel diseases and viruses that may cause epidemics and pandemics remain policy exclusions: a position that is unlikely to change given the considerable risks this would entail**. Despite the goodwill shown during the COVID-19 pandemic and the development of new products and riders incorporating coverage for the virus, the insurance industry remains cautious in accepting liability for future novel diseases. This is because the severity of any future outbreak is unpredictable; new viruses or diseases may be more contagious, require greater treatment support for longer periods of time, lead to more serious outcomes or mutate with greater speed leading to cycles of infection and reinfection. For this reason, there is little appetite to include coverage for novel viruses or diseases, even among the most advanced insurance markets globally.

3.4.6 Diagnostic and Recommended Actions

179.   **The payor hierarchy between government, PhilHealth and private insurers in relation to epidemics and pandemics should be established.** If private insurers had greater knowledge of the benefits and coverage that government—either directly or via PhilHealth—would offer for future epidemics and pandemics, it would be easier to understand potential gaps and develop complimentary products. Private insurers globally are assessing the potential for future health insurance epidemic and pandemic products, particularly that work in tandem with publicly funded systems, so there is scope for the Philippines to do the same.

The payor hierarchy between government, PhilHealth, and private insurers in relation to epidemics and pandemics should be established.

180.   **The availability of health products complementing the UHC is large albeit aimed to be consumed by the more affluent population.** Since the UHC has developed into a comprehensive health coverage the existing private sector health products appear to be sufficient and affordable to the targeted economic class.

No recommendation needed.

3.4.7 Business Interruption Insurance

181.   **Business interruption or loss of income insurance is offered in the Philippines, but uptake is limited.** The standard product is linked to fire or property insurance and the payout under the business interruption policy requires physical damage of the property. Pandemic-related business interruption or loss of income is thus not covered. The public awareness of the benefits of business interruption is low and, at the same time, the underwriting of business interruption or loss of income requires transparency in the income of the insurers, which is a challenge in most developing countries.

182.   **Significant business revenue losses have drawn attention to business interruption insurance globally.** COVID-19 pandemic measures including lockdowns, limiting size of gatherings, travel restrictions, etc., led to about $1.7 trillion in revenue losses as estimated by the OECD (Figure 15). The revenue losses in 2020 for the Philippines are estimated by the Japan Center for Economic Research to be around $43 billion. Absent risk transfer instruments, and to protect massive business closures and liquidations as well as exploding unemployment rates, governments globally responded with significant financial support programs. The need for business interruption or loss of income insurance has drawn public attention.

3.4.8 Diagnostic and Recommended Actions

183.   **Lack of pandemics risk diversification and the low demand for business interruption insurance need to be overcome for a viable risk transfer solution.** Having highlighted the difficulties to providing coverage for the financial impact of pandemics, this does not rule out the provision of small-scale selected private market coverage by limiting the degree of risk transfer and the number of businesses covered. Such availability of the product would benefit the Philippines.

Initially the government rather than individual companies should consider acquiring multi-year coverage for business interruption that includes pandemics/epidemics (paras. 182–183).

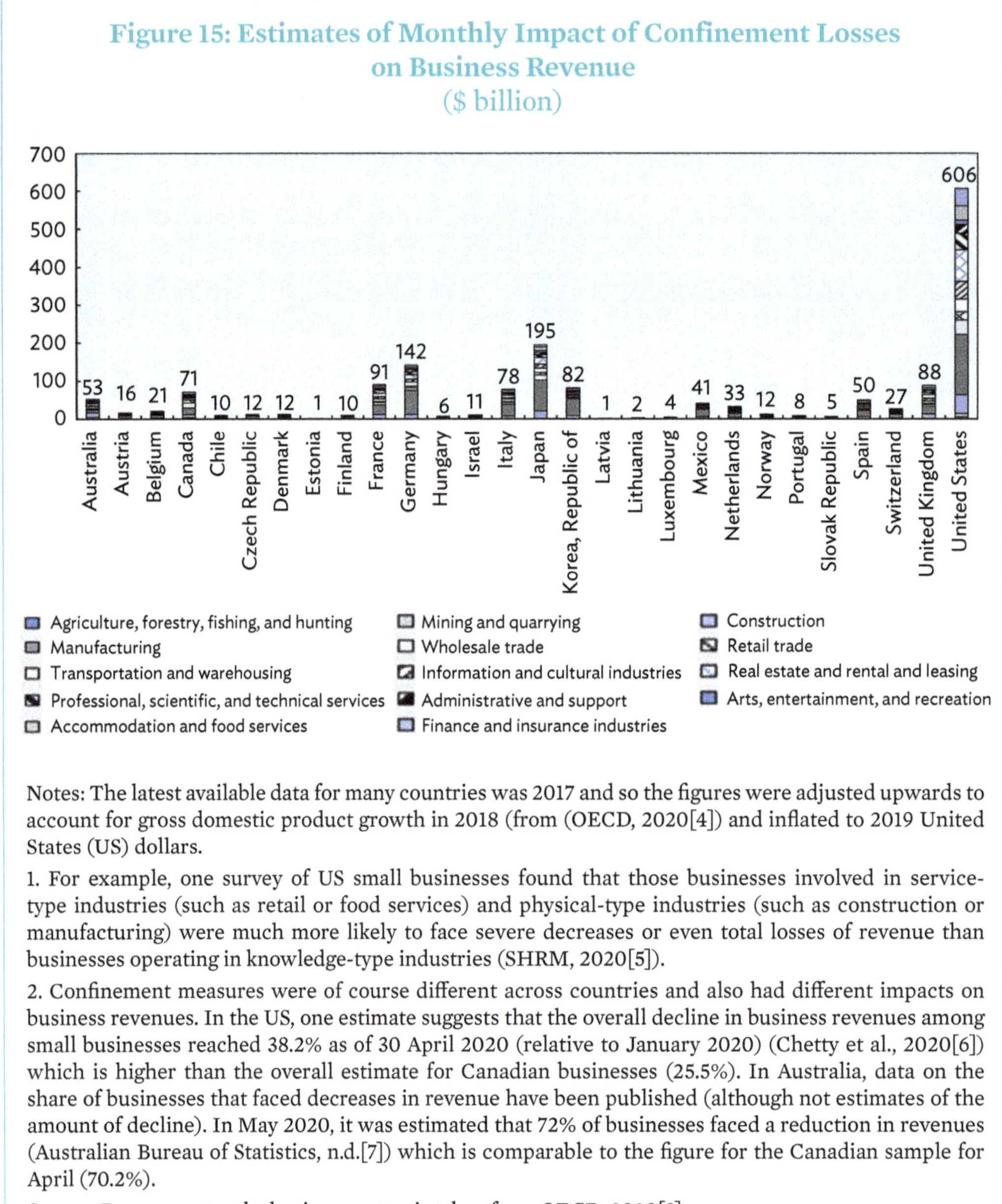

Figure 15: Estimates of Monthly Impact of Confinement Losses on Business Revenue
($ billion)

Notes: The latest available data for many countries was 2017 and so the figures were adjusted upwards to account for gross domestic product growth in 2018 (from (OECD, 2020[4]) and inflated to 2019 United States (US) dollars.

1. For example, one survey of US small businesses found that those businesses involved in service-type industries (such as retail or food services) and physical-type industries (such as construction or manufacturing) were much more likely to face severe decreases or even total losses of revenue than businesses operating in knowledge-type industries (SHRM, 2020[5]).

2. Confinement measures were of course different across countries and also had different impacts on business revenues. In the US, one estimate suggests that the overall decline in business revenues among small businesses reached 38.2% as of 30 April 2020 (relative to January 2020) (Chetty et al., 2020[6]) which is higher than the overall estimate for Canadian businesses (25.5%). In Australia, data on the share of businesses that faced decreases in revenue have been published (although not estimates of the amount of decline). In May 2020, it was estimated that 72% of businesses faced a reduction in revenues (Australian Bureau of Statistics, n.d.[7]) which is comparable to the figure for the Canadian sample for April (70.2%).

Source: Data on output by business sector is taken from OECD, 2020[3].

3.4.9 Travel Insurance

184.   **Standard travel insurance policies exclude epidemics and pandemics risk.** One of the main reasons for this exclusion is the fact that when epidemics and pandemics are declared, they become a known event, contradicting the central principle of insurance that it should cover events that have not occurred or do not have a 100% probability of occurrence in the duration of the policy. However, in the case of COVID-19 pandemic the requirement in several countries to have travel insurance policy that covers COVID-19 related costs has led insurance companies to design such a product. This has been possible due to the strict conditions for travel, like full vaccination, negative testing and quarantine.

3.4.10 Diagnostic and Recommended Actions

185.   **Due to the risen awareness of epidemics and pandemics risk, future travel insurance policies will have to include undeclared epidemics and pandemics in their coverage.** This will only be possible at an affordable price if similar conditions that mitigate the risk to travel are due to COVID-19, i.e., full vaccination, negative testing, etc.

Require future travel insurance policies to include undeclared epidemics and pandemics in their coverage.

3.4.11 Capital Markets Supporting Disaster Risk Transfer Instruments

186.   **The Philippine Stock Exchange, Inc. (PSE) has been developing but remains small compared with the Singapore Stock Exchange.** The PSE was created in 1992 and hosts around 275 listed companies The PSE index is composed of 30 companies including Alliance Global Group as the only company of the insurance sector. PSE is supervised by the Securities and Exchange Commission. Significant progress has been made since 2016 in efforts to develop the money and government securities markets, setting pricing benchmarks for corporate securities as indicated in the 2019 World Bank Financial Sector Assessment Program (Box 6).

Box 6: 2019 Financial Sector Assessment Program Key Findings
on the Development of the Capital Markets Since 2016

Significant progress has been achieved since 2016 in developing the short-end of the yield curve through increased issuance, initiating a predictable and transparent issuance policy, enhancing the primary dealer network, and adopting a new valuation methodology and trading platform for benchmark bonds. Emphasis has also been placed on stimulating the secondary markets by developing the repo markets for government securities. The initiatives undertaken represent a prudent and well-structured approach toward developing capital markets and should be continued. The authorities should prioritize: (i) resolving documentation issues related to the global master repurchase agreement, whereby foreign banks would prefer English law; (ii) broadening the repo market to include non-bank financial institutions that are not government securities eligible dealers, as they are already significant holders and secondary market traders; (iii) introducing anonymous trading in relation to the repo market, as the current system impacts the price and acts as a potential deterrent for market entrants; and (iv) regularize buybacks and exchanges of existing government securities.

Source: World Bank. Philippines Financial Sector Assessment. p. 33. https://openknowledge.worldbank.org/bitstream/handle/10986/36191/Philippines-Financial-Sector-Assessment.pdf?sequence=1&isAllowed=y.

187.   **Government interest is strong to continue developing the capital markets (Philippine News Agency 2021).** A significant Capital Market Development Act proposal that seeks to establish private retirement and pension schemes and the simplification of the taxation of passive income and financial services and transactions are in discussion, supported by the Department of Finance and the Capital Market Development Council.[85] In line with the support for deepening and further developing the capital markets that is within the ADB policy loan program (ADB 2020b), several recent improvements have been implemented:

- The Philippine Dealing System launched an electronic Securities Issues Portal
- Introduction of mobile software such as bonds.
- Availability of local and overseas bank apps, which have allowed the Bureau of the Treasury to widen its reach to individual investors in offering government bonds.

3.4.12 Diagnostic and Recommended Actions

188.   **The issuance and trading of insurance linked securities like CAT bonds issued for the benefit of the government will remain outside the PSE in the medium-term.** The 3-year CAT bonds issued in 2019 with the support of the World Bank were traded in the Singapore Stock Exchange, which is seeking to develop a regional CAT bond issuance platform. The further development of the domestic capital market will certainly create new funding opportunities for the economy but given the closeness to sophisticated exchanges, like Singapore and Hong Kong, China exchanges, the placement of insurance-linked securities probably will remain traded in those exchanges in the near term.

Develop further the PSE to become an efficient platform to issue the needed insurance-linked securities for the financing of disasters.

3.5 Social Protection Policy

3.5.1 Government Programs

189.   **GSIS and the Social Security System (SSS) are the two premier agencies responsible for social security in the Philippines.** Both GSIS and SSS practically carry their entire risks through in-house funds and there is no recourse to commercial reinsurance. Since both offer defined contribution programs, the risk is absorbed by the funds created from member contributions.

---

[85] Members of the Capital Market Development Council: The Secretary of Finance is chairman and the co-chairmen are the chairman of the Securities and Exchange Commission and a nominee from the Financial Executives Institute of Philippines (Finex), representing the private sector. On the government side, additionally a member of the Monetary Board of the Bangko Sentral ng Pilipinas, the Insurance Commissioner, the National Treasurer and the executive director of the Bureau of Local Government Finance. On the private sector side, the presidents or head executives of the Philippine Stock Exchange, the Philippine Depository and Trust Corp., the Philippine Dealing and Exchange Corp., the Bankers Association of Philippines, the Investment House Association of Philippines, the Fund Management Association of Philippines, the Philippine Association of Securities Brokers and Dealers Inc. and, of course, Finex.

190.   **The SSS offers coverage to formal sector employees and workers as well as to self-employed persons and those working in the unorganized sector, on a contributory basis.** The coverage for formal sector is compulsory while for others it is voluntary. The contributory schemes of SSS offer benefits ranging from sickness, maternity, retirement, death, disability as well as unemployment. Loans are also provided to eligible members under prescribed circumstances including following disasters. In 2019, SSS collected ₱182.00 billion toward contributions from members and serviced almost 2.5 million pensioners.[86] During January to September 2021, the SSS released ₱869.99 million worth of unemployment benefits to over 67,000 members (SSS 2021).

191.   **GSIS, on the other hand, performs the function of insuring government employees as well as its assets.** The benefit packages of GSIS are like that of the SSS, which includes almost 10 lines of insurance.

192.   **The Philippines has a Social Protection Framework that aims to reduce poverty and vulnerabilities.** The SSS and GSIS are tasked to respond on managing risks on diseases and ill-health, unemployment and underemployment, disasters triggered by natural hazards, and land and housing insecurity:

193.   **The Indigent Program of the Philippine Health Insurance Corporation (PhilHealth).** The Department of Social Welfare and Development identifies poor families who are part of the Pantawid Pamilyang Pilipino Program and Modified Cash Transfer and categorizes indigent members. Indigent members are those without visible means of income, or whose income is insufficient for family subsistence, as identified by the Department of Social Welfare and Development, based on specific criteria.[87] The premium contribution of indigent members are subsidized by the national government at around ₱2,400 ($47) per year. [88]

- **The Social Pension for Indigent Senior Citizens.** Starting 2022, senior citizens beneficiaries receive their ₱500 ($10) monthly stipend on a quarterly basis pursuant to the Department of Social Welfare and Development Memorandum Circular No. 16. This serves as an additional government assistance that aims to augment daily living subsistence and help manage medical needs of the senior citizen.[89]
- **The SSS Alkanssya Program for the Self-Employed in the Services Sector.** The program was established to accommodate cash inflows of informal sector workers. This enables daily SSS premium payments as low as ₱11.00 ($0.25). Members are registered as SSS self-employed members and required to meet the ₱3,000.00 ($67.38) monthly salary credit, a ₱330.00 ($7.41) monthly contribution. However, if they can afford it, workers can declare a higher monthly income than the required minimum and contribute a higher monthly salary credit (SSS 2015).
- Current members of this program include tricycle drivers, fisherfolk, home-based workers, women entrepreneurs, market vendors, golf club workers, jail inmates, materials recovery workers, and *barangay* (town) employees.

---

[86] Social Security System 2019 Annual Report.
[87] See PhilHealth Indigent Members, https://www.philhealth.gov.ph/members/indigent/.
[88] PhilHealth Circular No. 2017–0010 https://www.philhealth.gov.ph/circulars/2017/circ2017–0010.pdf.
[89] Department of Social Welfare Development, Memorandum Circular No. 16. Series of 2021 https://www.dswd.gov.ph/issuances/MCs/MC_2021–016.pdf.

- **SSS Contribution Subsidy Program**. This program targets farmers and agricultural workers in organized groups, aiming to tap agriculture-related government agencies or even private entities in shouldering a portion of the farmers' SSS premiums. Under this subsidy program, the farmer-member shoulders 50% of the SSS contribution amount and the government or private entity matches this with a corresponding 50% share. The premium share matched may be taken from foreign funding or grants or from the operating budget of the government agency (SSS 2015).

3.5.2 Private Sector Insurance Involvement in the Low-Income Population Protection

194.   **Private sector involvement in low-income population protection complements the government social security system.** While the government offers a basic level of social security through GSIS and SSS, the private sector offers broader social protection to communities and households through market-based insurance and protection products. The insurance industry provides complementary products covering life, health, assets, and disaster insurance products to low-income households, albeit with limited response by the low-income population.

195.   **Microinsurance plays a crucial role in offering market-based social protection products to low-income households in the Philippines.** For over 15 years, the Insurance Commission has been striving to promote microinsurance through enabling regulation, making the Philippines one of the first countries to develop microinsurance. In 2006, the Insurance Commission issued a ground-breaking circular that defined microinsurance, identified the essential features of a microinsurance policy, and lowered the initial guarantee fund requirements of an MBA wholly engaged in microinsurance (ADB 2017). This initiative has been regularly followed by further regulatory circulars. As a result, the microinsurance sector in the Philippines is driven by three sets of players: microinsurance mutual benefit associations (MI-MBAs), commercial insurers, and cooperative insurance societies (Box 7). While MI-MBAs enjoy a lower capital requirement, cooperative insurance societies and commercial insurers are treated at par in terms of capital requirement.

Box 7: Regulation of Cooperatives

Cooperatives in the Philippines are regulated by the Cooperative Code of the Philippines 1990. The Cooperative Development Authority is entrusted with the regulation and development of the cooperative sector in the country. In 2018, there were 18,065 operational cooperatives with a total of 10.7 million members in the country out of which 678 were agricultural cooperatives and 2,713 credit cooperatives. Under Filipino law, cooperatives and their members enjoy significant tax benefits.

Source: Cooperative Development Authority. 2018. State of the Cooperative Movement, Annual Report.

196.   **In general, the microinsurance sector has had decent growth over the past few years.** Out of 35 MBAs in 2019, 23 were MI-MBAs. This number was 34 versus 22 respectively in 2015. The number of members of MBAs increased from 4.64 million in 2015 to 10.8 million in 2019.[90] Even during the COVID-19 pandemic, lives covered under microinsurance increased decently, with MBAs continuing to take the largest market share in terms of estimated number of individuals covered by microinsurance. MBAs covered 27.43 million lives as of second quarter (Q2) of 2020—69.16% of total market share. In microinsurance premium production, MBAs had contributed ₱2.08 billion out of ₱3.54 billion as of Q2 2020—a share of 58.78% (Insurance Commission 2020). Premium collection by MBAs grew almost 60% from ₱6.9 billion in 2015 to ₱11 billion in 2019. In the same period, the share of premium collections by MI-MBAs increased from 36% to 44% of the total premium collected by all MBAs (footnote 90).

However, premium collections declined during the COVID-19 pandemic. Premium collection declined in Q2–2020—by 17.84% year-on-year for MBAs and 14.5% for life insurers—as both extended premium payment grace periods during the pandemic.[91] The nonlife insurance sector posted a 24.79% decrease in the number of lives covered year-on-year, from 5.04 million in Q2–2019 to 3.79 million. Microinsurance premium production declined 14.48% from ₱473.74 million as of Q2–2019 to ₱405.13 million as of Q2 2020 (footnote 91).

197.   **In the Philippines, MBAs are allowed to underwrite life insurance, subject to certain limits on sum insured and premium.** Since microfinance has been the major driver behind MBAs, credit life insurance products attached to microfinance loans form a large part of the MBA insurance portfolio. MBAs also offer a voluntary basic life insurance product that covers members and family, if opted to do so. One part of the contribution collected from the members goes toward a basic life insurance product risk premium, while the other is appropriated toward members' equity. MI-MBAs also offer add-on covers such as permanent total disability and hospitalization arising out a motor vehicle accident. In addition, MI-MBAs offer Golden Life Insurance Plans to people aged 70 years and older, who are beyond the insurable age of credit life insurance products and basic life insurance product. The sum insured for a Golden Life Insurance Plan is generally restricted to ₱50,000. Cooperative insurers like 1 Cooperative Insurance System of the Philippines (1CISP) and CLIMBS Life and General Insurance Cooperative also offer long term savings linked life insurance that offers protection plus a guaranteed sum on maturity.

198.   **Cooperatives and nonlife insurers provide nonlife products.** Non-life insurers provide indemnity-based insurance for dwellings, shops, and stocks.

199.   Several cooperative insurance societies provide nonlife products to its members:

- 1CISP in collaboration with German Agency for International Cooperation is also contemplating a parametric business interruption product called Nego Seguro Plus for small businesses that will pay a fixed sum of money upon a typhoon of the specified intensity hitting the area in which the insured business is located.

---

[90] Insurance Commission. Key Statistical Data 2015–2019.
[91] IC statistics 2020.

1CISP also launched a fintech platform called DigiCoop that allows online transactions like lending, money transfer, policy issuance, premium and claims payments.

- CARD-Pioneer Microinsurance offers an indemnity-based AgriMicro crop insurance products that covers specified perils including disasters. Being an indemnity policy, it involves farm level loss assessment.

200. **PCIC is mandated to provide strongly subsidized insurance protection to the low-income farmers.** To provide protection to the agricultural households that are among the most vulnerable segment of the population and highly exposed to disasters (Box 8), PCIC offers full premium subsidy to beneficiaries in the RSBSA for rice and corn crops while partial subsides are offered to other farmers under various programs of Department of Agriculture. (For more details on PCIC and agriculture insurance, see Section 3.3.5).

Box 8: Poverty Incidence Among Agriculture Households

A poverty study undertaken in 2012 estimated that the poverty incidence among agricultural households (57%) is thrice than that of the nonagricultural households (17%). More so, food poverty or subsistence incidence among agricultural households is about five times greater than those among nonagricultural households. Ironically, agricultural food producers are food-poor.[a] According to another study "shocks are the major factor contributing to impoverishment and remaining in poverty."[b] Given that the Philippines is highly prone to disasters, famers are all the more vulnerable to financial shocks resulting from disasters.

[a] PIDS. Poverty and Agriculture in Philippines: Trends in Income Poverty and Distribution. *PIDS Discussion Paper Series* No. 2012–09.
[b] International Fund for Agricultural Development. 2011. Rural Poverty Report: New Realities, New Challenges, New Opportunities for Tomorrow's Generation.

3.5.3 Public Health Interaction with the Private Sector Providers

201. **Health protection, like in most countries, is a combination of social and private insurance.** Health protection is provided by PhilHealth through the state's social health insurance the National Health Insurance Programme as well as by the private sector through HMOs and insurers. To make sure the plans complement each other, the priority of the policies has been established by the government. Accordingly, private plans must be exhausted before National Health Insurance Programme benefits can be used. However, in the case of COVID-19, the government ruled that PhilHealth was responsible for the medical expenses in the first instance (see sections 3.2.3, 3.3.5, and 3.4.5 for more details on the health system and the role of the different actors).

3.5.4 Diagnostic and Recommended Actions

202. **Most of the MI-MBAs are part of a microfinance institution (MFI) and thus can only serve MFI members.** The MBA sector in the Philippines emerged as an offshoot of the microfinance sector. Even today most MI-MBAs are part of an MFI. This has restricted the outreach of MBAs to MFI clients only.

While MFI sector itself in the Philippines is huge, MI-MBAs need to develop strategies to tap the low-income market beyond the micro-credit clientele. *They should use the special regulatory dispensation available to them to broaden their clientele base and thereby reap the benefits of better diversification.*

203.   **All insurance products currently attract VAT as well as stamp duty.** In the case of nonlife policies, including microinsurance products, the combined financial incidence of VAT and stamp charges goes up to more than 20% of the premium.[92]

To make nonlife microinsurance products affordable among the low-income households, exemptions from both these taxes are needed. *All insurance products, life as well as nonlife filed as microinsurance products should be given a blanket exemption from VAT and stamp duty. Because of the small ticket size of the policy, the revenue implications of such an exemption on the state would be insignificant, but the products, especially nonlife microinsurance products will become much cheaper for the consumers.*

204.   **Currently, the microinsurance product basket is limited to life insurance.** When disasters strike, small businesses are worst affected, in material damage and temporary business interruption. Pandemics like COVID-19 also cause temporary loss of livelihoods and diagnostic and treatment related costs. Covering these risks through standard indemnity policies is costly due to the need for individual loss assessment.

The insurance industry needs to develop small ticket parametric products covering health and livelihood risks of the poor households, especially for disaster, epidemic and pandemic cover. *In addition to indemnity policies, parametric products that pay a fixed amount of money per day upon the occurrence of qualifying natural hazard events or the diagnosis of a specified ailment/infection like COVID-19, should be developed. The administration of these products is easier and can provide immediate liquidity to the population in the affected area. The number of days for which the lump sum amount is paid can also be stipulated based on the severity of the typhoon or type of disease. As and when the volumes go up, such products can be ceded into the PCIF if it is set up*

3.6   Unlicensed Competition

205.   **The Insurance Commission reports companies operating without a license in its webpage.** The last reported activity dates from 2017, showing that unlicensed activity is under control. All entities identified as operating without a license are small, pre-need companies, thus the Insurance Commission is following up on unlicensed activity.

---

[92] Manila Times (2022). See Taxes On Nonlife Insurance Policies.

Conclusions

4.1 Rating Summary and Recommended Main Actions

206. **The ideal enabling environment for DRF coincides with the achievable scenario for the Philippines and the gap analysis of the current enabling environment has therefore been carried out against the ideal scenario.** Responses from stakeholders regarding the realistically achieved enabling environment for DRF for the Philippines were more by way of providing additional solutions for achieving the ideal scenario, rather than identifying limitations that would hinder realization of the ideal. The figure presenting the ratings thus shows only the current situation versus the ideal enabling environment (Figure 16).

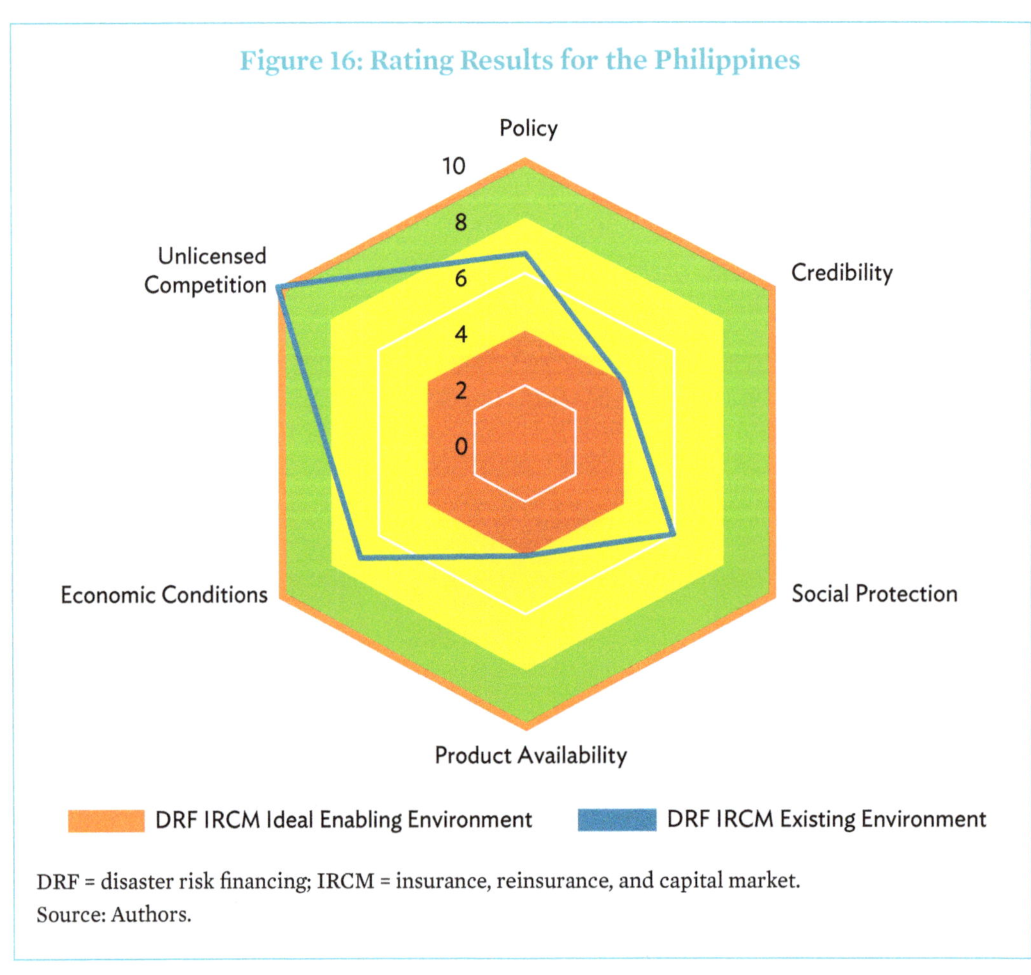

Figure 16: Rating Results for the Philippines

DRF = disaster risk financing; IRCM = insurance, reinsurance, and capital market.
Source: Authors.

4.1.1   Economic and Other Preconditions

207.   **The rating is in the yellow zone, implying need for action (section 3.1).**

Main gaps identified:

- Historic and modeled risk data, both at the aggregated and granular level necessary to develop and design suitable DRF instruments is not readily available.
- The fixed asset registers do not include valuation at replacement value.
- A shortage of physicians and nurses and insufficient supply of resources across the country limit the respond to pandemics.
- While important pandemic and epidemic data collection and monitoring arrangements are in place, there are areas of further strengthening.
- A dedicated portal for agriculture where all data available can be integrated does not exist.
- Substantial investments in agritech will be needed to make agriculture to withstand disasters efficiently and enhance its insurability.

Main recommended actions to close the gaps:

- Continue improving the availability of and access to disaster risk data, including ensuring wide availability of robust, granular data by government agencies, the insurance sector, and other nonsovereign users.
- Fixed asset registers, once established, should be reviewed every 3 to 5 years, including to update the valuation of assets if required. Valuation data should include replacement value.
- Strengthen financing in the healthcare infrastructure to ensure adequate levels of health resilience to epidemics and pandemics, especially in the more remote areas of the country.
- Evolve integrated health surveillance systems through a collaborative effort between the government, the WHO, and other parties involved in the Joint External Evaluation process.
- Continue providing anonymized, aggregate data on infectious disease outbreaks publicly available.
- Make data on the total cost of health delivery available.
- Launch a dedicated portal for agriculture, where all data available can be integrated and made accessible to registered government and private sector entities.
- Channel substantial investments in agritech to increase efficiency, predictability and thus reduce agriculture risk to a level where insurance becomes sustainable and affordable.

4.1.2 Government Policy

208.   **The rating is in the yellow zone, implying need for action (sections 2 and 3.2).**

Main gaps identified:

- The appropriation for the NDRRMF is currently determined on an annual basis compromising long term planning.
- The local special purpose disaster funds have not been fully used in the past.
- The National Tax Allotment methodology for Local Government does not recognize the impact of disaster risks other than at a general level of population and land area.
- A centralized register of disaster related contingent liabilities is not fully operational.
- Financing for disaster risk management exists in potentially significant budget sums beyond the main program and the NDRRMF in mainstream, regular budgets.
- The 2019 Open Budget Survey identified a gap in registration, valuation and disclosure of disaster-related fiscal risks and contingent liabilities.
- The use of parametric catastrophe risk insurance needs improvements.
- There is lack of insurance or under-insurance of assets under local government units' control that are critical to disaster responses including hospitals and care institutions.
- With regard to the COVID-19 pandemic, there was insufficient clarity regarding the role of the national government versus the role of local government units in responding to the pandemic, particularly in relation to planning and financing the health response.
- There is no clear responsibility on PhilHealth's role in financing the direct healthcare costs arising from epidemics and pandemics.
- PhilHealth does not use reinsurance to mitigate the government's exposure to direct healthcare costs relating to epidemics and pandemics.
- There is a need to clarify the provider payment mechanisms that PhilHealth can use during times of disaster.
- COVID-19 pandemic has led to delays in planned UHC reforms.
- Health security and preparedness need strengthening in line with International Health Regulations standards, among other benefits, to reduce pandemic and epidemic risk to encourage economically viable insurance to emerge.
- Zoonotic disease management, as noted, is underdeveloped, which is likely to make the risk of economic losses from epidemics and pandemics high.
- Pandemic and epidemic risk transfer instruments are not part of the sovereign DRF instruments.
- The agriculture sector depends heavily on subsidies, that while supporting the use of insurance, also create certain behavioral anomalies disincentivizing individual risk management.

Main recommended actions to close the gaps:

- Consider automatic appropriation status for funding the national government NDRRMF based on a percentage of GDP or other formula to be determined.
- Incentivize local government units to spend more on proactive DRR and risk transfer solutions and also to consider arrangements for pooling quick response funds.

- Review the Local Disaster Risk Reduction and Management Fund special trust fund and carry forward rules to be more risk-led rather than universally applied.
- Consider a separation of the local special purpose disaster funds for DRR and DRF
- Revise the National Tax Allotment methodology to take account of relative climate change and disaster risk.
- Establish a central register of contingent liabilities that identifies and values disaster, epidemic and pandemic related fiscal risks.
- The government should consider conducting a Disaster Risk Management Public Expenditure and Institutional Review.
- Investigate the use of temporary expenditure markers to identify the costs of disasters at event level.
- Improve the use of parametric catastrophe risk insurance:
- The intermediation of parametric insurance should be cost efficient and the process to disburse claims payments to the ultimate beneficiaries should be timely and based on a pre-established methodology.
- If the payout trigger is related to a model loss calculation, such a model should be fully understood by the government, including the ultimate beneficiaries, to determine the usefulness of the disaster financing instrument.
- Incentivize local government units to take out appropriate insurance to cover public assets including medical infrastructure.
- Include additional details in the policy on health system decentralization on roles and responsibilities, especially with respect to financing disasters, epidemics, and pandemics.
- Clarify the role and responsibility of PhilHealth in financing direct healthcare costs arising from epidemics and pandemics.
- Explore the use of reinsurance to protect PhilHealth in case of a pandemic or epidemic.
- Reassess the provider payment mechanisms that PhilHealth can use during times of disaster, epidemic and pandemic.
- Maintain progress toward comprehensive UHC that reduces existing health protection gaps, bolsters health spending and enhances overall system resilience.
- Utilize the findings of the 2018 Joint External Evaluation to move toward full compliance with the International Health Regulations standards.
- Improve the functionality and effectiveness of local government units in zoonosis control and prevention, heighten public health literacy, and enhance public health campaigns.
- Acquire sovereign pandemic and epidemic risk transfer instruments to provide funds at the points of need when available.
- Use subsidies to encourage and enhance farmers' risk management skills to reduce the cost of insurance.

4.1.3 Credibility in the Insurance Sector and the Capital Markets

209.   The rating is on the red/yellow zone boundary, implying urgent need for action (section 3.3).

Main gaps identified:

- The operational independence of the Insurance Commission suffers from certain limitations.
- Planned additional increment on the already very high minimal capital could reduce competition and increase insurance costs.
- Key participants in the provision of insurance to the public—namely, Philippine Crop Insurance Corporation, Government Service Insurance System and PhilHealth—lag compliance with insurance technical practices and need effective insurance supervision
- Private sector involvement in agriculture insurance remains very limited despite efforts to encourage it.
- A huge agricultural sector protection gap still exists despite the good work done by PCIC in covering the poor farmers of the Philippines.

Main recommended actions to close the gaps:

- Improve the operational independence of the Insurance Commission accompanied by measures to increase its formal accountability to the government.
- Maintain the minimum capital requirements at the current level of ₱900 million but strengthen the monitoring and compliance with the RBC2 requirements.
- Assign the insurance supervision of PCIC, GSIS and PhilHealth to the Insurance Commission.
- Continue encouraging private sector involvement in agriculture insurance, especially for commercial farmers and nontraditional crops.
- Promote public–private partnerships for agriculture insurance to close the protection gap in agriculture.

4.1.4 Products

210.   The rating is in the red/yellow border zone, implying urgent need for action (section 3.4).

Main gaps identified:

- Access to asset and property disaster risk insurance is limited in certain areas of the country.
- Agriculture involves a wide range of risks with varying frequency and severity, making necessary risk sharing arrangements between all stakeholders that are not in place.
- Agriculture reinsurance is not readily available at an affordable price.
- The expansion of agriculture insurance will require a larger basket of products than is currently available.
- The payor hierarchy between the government, PhilHealth and private insurers in relation to epidemics and pandemics remains unclear.
- Significant business revenue losses have remained uninsured.
- Travel insurance excludes pandemics.
- The issuance and trading of insurance linked securities, like CAT bonds issued for the benefit of the government, remain outside the PSE.

- Provide technically sound catastrophe risk insurance through the PCIF, if established, to any part of the country where residential and other buildings are officially allowed to be built.
- Encourage through smart subsidies a risk-sharing arrangement between farmers' cooperatives and commercial insurers.
- Establish an agriculture insurance pool to fill the gap between domestic and international reinsurance markets, allowing higher domestic agriculture risk retention.
- Include on agriculture product innovations value-added services.
- Establish the payor hierarchy between government, PhilHealth and private insurers in relation to epidemic and pandemic claims.
- Consider multiyear business interruption insurance for the government.
- Require future travel insurance policies to include undeclared epidemics and pandemics in their coverage.
- Develop further the PSE to become an efficient platform to issue the needed insurance-linked securities for the financing of disasters.

4.1.5 Social Protection

211.   The rating is in the yellow zone, implying a need for action (section 3.5).

Main gaps identified:

- Most of the MI-MBA are part of an MFI and thus only serve MFI members.
- All insurance products currently attract VAT as well as stamp duty
- Currently the microinsurance product basket is limited to life insurance products.

Main recommended actions to close the gaps:

- MI-MBAs should develop strategies of tapping the low-income market beyond the micro-credit clientele.
- Consider granting full tax exemption to nonlife microinsurance products
- Develop small ticket parametric products covering health and livelihood risks of the poor households, especially risks linked to disasters, epidemics and pandemics

4.1.6 Unlicensed Competition

212.   The rating is in the green zone, implying no action is needed (section 3.6).

Main gaps identified:

- No gaps were identified

Main recommended actions to close the gaps:

- None

Appendix: Property Insurance Law of Public Assets

REPUBLIC ACT NO. 656:[93] **An Act to Create and Establish a "Property Insurance Fund" and to Provide for its Administration and for other Purposes"**

Section 1. This Act shall be known as the "Property Insurance Law."

Section 2. To indemnify or compensate the Government as defined in this Act for any damage to, or loss of, its properties due to fire, earthquake, storm, or other casualty there is hereby established the "Property Insurance Fund," which shall consist of all moneys resulting from the liquidation of the insurance constituted in section three hundred forty of the Revised Administrative Code and from premiums and other incomes.

Section 3. For the effectuation of the purpose of this Act, the administration of the Fund is hereby placed under the Government Service Insurance System with powers and authority to reinsure with private insurance companies under such terms and conditions that may be mutually agreed upon any excess risk it may deem advisable; to prescribe necessary rules and regulations, including such incidental powers as are necessary for its operation; and to appoint personnel, who are certified as eligible by the Civil Service, prescribe their duties, and fix their remuneration. Section fifteen of Commonwealth Act Numbered One hundred eighty-six shall not be applicable to the personnel of the Fund.

Section 4. Definitions (For this Act)

 (a) "System" means the Government Service Insurance System created under Commonwealth Act Numbered One hundred and eighty-six.

 (b) "Fund" means the "Property Insurance Fund" created under this Act.

 (c) "Property" includes vessels and craft, motor vehicles, machineries, permanent buildings, properties stored therein, or in buildings rented by the Government, or properties in transit.

 (d) The word "Government" as used in this Act refers to the National, provincial, city, or municipal government, agency, commission, board or enterprises owned or controlled by the Government.

 (e) "Disposable Surplus" means the amount left after the necessary insurance reserves and other reserves have been set aside together with the expenses incidental to the administration of the Fund.

---

[93] RA 656: Property Insurance Fund. https://www.gsis.gov.ph/general-insurance/general-insurance-mandate/republic-act-no-656/.

Section 5. Every government, except a municipal government below first class, is hereby required to insure its properties, with the Fund against any insurable risk herein provided and pay the premiums thereon, which, however, shall not exceed the premiums charged by private insurance companies: Provided, however, That the System reserves the right to disapprove the whole or a portion of the amount of insurance applied for: Provided, further, That such property or part thereof as may not be insurable or acceptable for insurance may be insured with any private insurance company. A municipal government below first class may upon application insure its properties in the Fund under such rules and regulations as the System may prescribe.

Section 6. Collection and payment of premiums.—(a) In accordance with such rules and regulations as the System may prescribe under section three of this Act, the premiums on insurance under section five hereof shall be paid in advance to the System by the government concerned. (b) Penalties.—Any cashier, treasurer, or any government official responsible for the collection and/or remittance of the premiums hereinabove prescribed, who refuses or habitually neglects to comply with the instructions of the System and to collect or accept payments of the said premiums, issue receipts therefor, and/or remit the same within the time prescribed by the System, shall be held liable for the payment of said premiums and shall pay to the System a fine of two per centum per month of said premiums from their due dates until received by the System.

Section 7. (a) There shall be collected, classified, analyzed, and kept under the custody of the System such statistical data as may be necessary for the proper determination of risks and rates of premiums for the issuance of properties herein defined. (b) The records and accounts of the Fund shall be kept separate and distinct from those of other funds of the System. (c) During the month of October of each year, the System shall submit to the President a report of the operations of the Fund during the preceding year.

Section 8. The Auditor General and the Government Corporate Counsel shall be the ex-officio auditor and legal adviser of the Fund, respectively. The Auditor General, or his authorized representative, shall submit to the System soon after the close of each fiscal year audited statements showing its financial condition and progress for the fiscal year just closed and the actuary shall likewise make an actuarial examination and valuation of the fund.

Section 9. Any disposable surplus that may result from the operation of this Fund once declared by the System shall be apportioned in accordance with the schedule approved by the System among the governments whose properties are insured in the Fund. The Government of the Republic of hereby guarantees the fulfillment of the obligations of the Fund when and as they shall become due.

Section 10. Upon approval of this Act, the System shall require the inventory of the property belonging to each government, determine the value thereof for purposes of insurance, and advise the government concerned of the total premiums each shall pay to the Fund. The government concerned shall, upon receipt of advice from the System, set aside from any savings in its appropriation the amount needed for such premiums and certify to the availability thereof, and the Auditor General or his duly authorized representative shall forthwith release the same to the credit of the System by means of a journal voucher drawn for the purpose.

Thereafter, the official concerned shall remit it to the System immediately: Provided, That the premiums corresponding to the insurance of properties belonging to an entity operated with a special fund shall be payable from said fund; otherwise, from the general fund.

Section 11. Each government as defined herein shall include in its annual appropriation the amount necessary to cover the premiums for the insurance of its properties during each fiscal period and remit the same immediately to the System as provided in section ten hereof.

Section 12. Chapter sixteen of the Revised Administrative Code, as amended, is hereby repealed, and all the present assets and liabilities of the Property Insurance Fund created thereunder shall be liquidated upon approval of this Act and turned over to the System. The Auditor General and the Commissioner of the Budget or their authorized representative shall carry out the provisions of this section: *Provided*, That the properties insured under the Property Insurance Fund shall continue to be governed by existing law until they shall have been insured anew with the fund under the provisions of this Act.

Section 13. This Act shall take effect on the first day of the third calendar month following that of its approval.

Approved, 16 June 1951.

References

Asian Development Bank (ADB). 2013. *Investing in Resilience: Ensuring a Disaster-Resistant Future.*

———. 2014. *Operational Plan for Integrated Disaster Risk Management, 2014–2020.*

———. 2017. *Assessment of Microinsurance as an Emerging Microfinance Service for the Poor. Updated January 2017.*

———. 2018. *Philippine City Disaster Insurance Pool: Pool Rationale and Design.*

———. 2020a. *COVID-19 Impact on International Migration, Remittances, and Recipient Households in Developing Asia.* ADB Briefs. https://www.adb.org/sites/default/files/publication/622796/covid-19-impact-migration-remittances-asia.pdf.

———. 2020b. *Philippines: Support to Capital Market-Generated Infrastructure Financing Program-Subprogram 1.*

———. 2020c. *Report and Recommendation of the President to the Board of Directors: Proposed Policy-Based Loan to the Philippines for the Disaster Resilience Improvement Program.*

———. 2020d. Technical Assistance for Strengthening the Enabling Environment for Disaster Risk Financing (Phase 2).

———. 2021a. *COVID-19 and the Finance Sector in Asia and the Pacific-Guidance Note.* https://www.adb.org/sites/default/files/institutional-document/761946/covid-19-finance-sector-asia-pacific-guidance-note.pdf.

———. 2021b. Philippine Economy Seen Recovering in 2021 and with Stronger Growth in 2022. News release. 28 April 2021. https://www.adb.org/news/philippine-economy-seen-recovering-2021-stronger-growth-2022-adb.

———. 2022a. *Asian Development Outlook 2022.* ADB. www.adb.org/sites/default/files/publication/784041/ado2022.pdf.

———. 2022b. COVID-19 and Overseas Filipino Workers–Return Migration an Reintegration into the Home Country—The Philippine Case. *ADB Southeast Asia Working Paper Series.* https://www.adb.org/sites/default/files/publication/767846/sewp-021-covid-19-ofws-return-migration-reintegration.pdf.

———. 2023. *Asian Development Outlook 2023.* ADB. https://www.adb.org/publications/asian-development-outlook-april-2023.

———. Forthcoming. *Assessing the Enabling Environment for Disaster Risk Financing (revised edition).*

ADB and the World Bank. 2017. *Assessing Financial Protection against Disasters: A Guidance Note on Conducting a Disaster Risk Finance Diagnostic.*

Agricultural Credit Policy Council. State of Agricultural Finance 2020. https://acpc.gov.ph/wp-content/uploads/POLICY-FORUM.pdf.

Bank for International Settlements. 2020. *Annual Report Economic Report 2020.* https://www.bis.org/publ/arpdf/ar2020e1.pdf.

Bangko Sentral ng Pilipinas (BSP). 2020. *COVID-19 Exit Strategies: How Do We Proceed? BPS Working Paper Series* No. 2020–01.

———. 2020. *Shifting Macroeconomic Landscape and the Limit of the BSP's Pandemic Response. BSP Working Paper Series* No. 2020–05. https://www.bsp.gov.ph/Media_And_Research/WPS/WPS202005.pdf.

BSP. BSP Governor Speech Philippines: Traversing the Path to Economic Recovery. 5 May 2021. https://www.bsp.gov.ph/SitePages/MediaAndResearch/SpeechesDisp.aspx?ItemId=796#:~:text=In%20response%20to%20the%20COVID,regulatory%20relief%20and%20forbearance%20measures.

Bloomberg. 2022. Covid Resilience Ranking. The Best and Worst Places to Be as World Enters Next COVID Phase. 29 June. https://www.bsp.gov.ph/Media_And_Research/WPS/WPS202001.pdf.

Bureau of the Treasury. National Government Outstanding Debt 2016–2021. https://www.treasury.gov.ph/wp-content/uploads/2022/02/Annual-Debt-1986–2021_web.pdf.

Center for Local and Constitutional Reform. 2020. What is Mandanas Ruling? 3 August. https://constitutionalreform.gov.ph/ufaqs/what-is-mandanas-ruling/.

S.A. Cole, G. Xavier, and J. Vickery. 2013. How Does Risk Management Influence Production Decisions? Evidence from a Field Experiment. *Policy Research Working Paper Series* 6546. The World Bank. https://ideas.repec.org/p/wbk/wbrwps/6546.html.

Cooperative Development Authority. 2018. *State of the Cooperative Movement, Annual Report.*

Development Budget Coordination Committee (DBCC). 2022. Fiscal Risks Statement 2022. The Department of Budget and Management. Fiscal-Risks-Statement-2022-for-Circulation.pdf.

Department of Budget and Management. 2020 People's Proposed Budget. https://www.dbm.gov.ph/images/pdffiles/2020-Peoples-Proposed-Budget.pdf.

Department of Budget and Management. 2021 People's Proposed Budget. p. 38. https://www.dbm.gov.ph/images/pdffiles/2021-Peoples-Proposed-Budget.pdf.

Department of Budget and Management. 2022 People's Proposed Budget. p. 59. https://www.dbm.gov.ph/images/pdffiles/2022-Peoples-Proposed-Budget.pdf.

Department of Budget and Management. 2020 Fiscal Risks Statement. https://www.dbm.gov.ph/wp-content/uploads/DBCC_MATTERS/FiscalRiskStatement/FY-2020-Fiscal-Risks-Statement.pdf.

Department of Budget Management. 2022. *Development Budget Coordination Committee. 2022 Fiscal Risk Statement.* https://www.dbm.gov.ph/wp-content/uploads/DBCC_MATTERS/FiscalRiskStatement/Fiscal-Risks-Statement-2022-for-Circulation.pdf.

Department of Budget and Management. Fiscal Risks Statement. https://www.dbm.gov.ph/index.php/dbcc-matters/dbcc-publication/fiscal-risk-statement.

Department of Budget and Management. Calamity and Quick Response Funds. https://www.dbm.gov.ph/index.php/programs-projects/calamity-and-quick-response-funds#1-what-is-calamity-fund.

Department of Budget and Management. Status of National Disaster Risk Reduction and Management Fund. https://www.dbm.gov.ph/index.php/programs-projects/status-of-national-disaster-risk-reduction-and-management-fund#2021.

Department of Finance (DOF). 2007. LGU Fiscal and Financial Profile Volume 1 – 2004. Local Government Bureau.

DOF. National Government Cash Budget % to GDP. https://www.dof.gov.ph/data/national-government-cash-budget/ (accessed 23 June 2022).

DOF. 2007. Bureau of Local Government Finance. LGU Fiscal and Financial Profile. https://blgf.gov.ph/wp-content/uploads/2015/08/SIE_CY_2004_Volume_1_new.pdf.

Department of Health. 2014. *Manual of Procedures for the Philippine Integrated Disease Surveillance and Response 3rd Edition.*

Department of Health. 2021. *DOH and Other Operating Units Committed to Resolving COA Findings; PHP 67.3 Billion COVID-19 Funds Mostly Resolved.* Press release, 17 August. *https://doh.gov.ph/press-release/DOH-AND-OTHER-OPERATING-UNITS-COMMITTED-TO-RESOLVING-COA-FINDINGS-PHP-67.3-BILLION-COVID-19-FUNDS-MOSTLY-RESOLVED.*

J. Elliot, M. Erbenova, A. Pancorbo. 2013. Operational Independence: Conceptual Framework Supervisory Independence. *IMF Working Paper* 09–2013.

Food and Agricultural Organization (FAO). 2022. COVID-19 Pandemic Impacts on Asia and the Pacific. https://www.fao.org/3/cb8594en/cb8594en.pdf.

D. Eckstein, V. Künzel, and L. Schäfer. 2021. *Global Climate Risk Index 2021.* German Watch. https://reliefweb.int/sites/reliefweb.int/files/resources/Global%20Climate%20Risk%20Index%202021_1_0.pdf.

Insurance Commission. 2018. *A Handbook on Microinsurance.* p. 17. https://micorner.insurance.gov.ph/microinsurance/wpcontent/uploads/2019/11/MicronInsurance-hand-book.pdf.

Insurance Commission. Statistics 2020.

International Budget Partnership. 2021. *Managing Covid Funds: The Accountability Gap.* https://internationalbudget.org/covid/wp-content/uploads/2021/05/Report_English-2.pdf.

International Federation of Accountants. *International Public Sector Accounting Standard. IPSAS 17 - Property, Plant and Equipment.* https://www.ifac.org/system/files/publications/files/ipsas-17-property-plant-2.pdf.

———. 2019. Case Study: Adoption of International Public Sector Accounting Standards in the Philippines. https://www.ifac.org/system/files/publications/files/Case-Study-Adoption-of-IPSAS-Philippines.pdf.

International Fund for Agriculture Development. 2011. *Rural Poverty Report: New Realities, New Challenges, New Opportunities for Tomorrow's Generation.*

International Monetary Fund. 1999. Contingent Government Liabilities—A Hidden Fiscal Risk. *Finance & Development*. 36 (1). https://www.imf.org/external/pubs/ft/fandd/1999/03/polackov.htm#:~:text=Implicit%20liabilities%20represent%20moral%20obligations,public%20expectations%20or%20political%20pressures.

International Monetary Fund (IMF). 2021. *IMF Country Report No 21/177—Executive Board Assessment*.

The Manila Times. 2022. Taxes on Nonlife Insurance Policies. 1 January.

International Cooperative and Mutual Insurance Federation. 2018. Mutual and Cooperative Microinsurance in Philippines: A Landscape Study.

Office of Civil Defense–Policy Development and Planning Service. 2020. *National Disaster Risk Reduction Plan 2020–2030*. Office of Civil Defense: https://ndrrmc.gov.ph/attachments/article/4147/NDRRMP-Pre-Publication-Copy-v2.pdf.

International Budget Partnership. 2019. Open Budget Survey. Rankings. https://www.internationalbudget.org/open-budget-survey/country-results/2019/philippines.

Organisation for Economic Co-operation and Development (OECD). 2021. Responding to the COVID-19 and Pandemic Protection Gap in Insurance. https://www.oecd.org/coronavirus/policy-responses/responding-to-the-covid-19-and-pandemic-protection-gap-in-insurance-35e74736/.

Philippine Crop Insurance Corporation. 2020 Annual Report. https://pcic.gov.ph/wp-content/uploads/2021/10/annual-2020-for-website-final.pdf.

Philippine Institute of Development Studies (PIDS). Poverty and Agriculture in Philippines: Trends in Income Poverty and Distribution. *PIDS Discussion Paper Series* No. 2012–09.

———. 2014. Review of Design and Implementation of the Agricultural Insurance Programs of the PCIC. *Discussion Paper Series* No. 2015–07.

———. Crop Insurance Program of the PCIC: Integrative Report from the Five Case Regions in the Philippines. *Discussion Paper Series* No. 2017–39.

Philippine News Agency. 2021. Capital Market to Lead PH Economic Recovery. 25 May. https://www.pna.gov.ph/articles/1141477.

Philippine Statistics Authority (PSA). 2015. Special Report - Highlights of the 2012 Census of Agriculture.

———. 2019. Agriculture Indicators System, Agricultural Credit.

———. 2020. Damages Due to Natural Extreme Events and Disasters, 28 October 2020. https://psa.gov.ph/content/damages-due-natural-extreme-events-and-disasters-amounted-php-463-billion.

———. 2020. Regional Compendium of Environment Statistics 2020.

———. 2021. Selected Statistics on Agriculture.

Republic of the Philippines, Congress of the Philippines, *Republic Act No 10121 An act strengthening the Philippine disaster risk reduction and management system, providing for the national disaster risk reduction and management framework and institutionalizing the national disaster risk reduction and management plan. May* 2010. https://www.officialgazette.gov.ph/2010/05/27/republic-act-no-10121/.

Republic of the Philippines, Office of the President, *Executive Order 10*, 2011.

Republic of the Philippines. Office of the President of the Philippines, *Executive Order 192 Transferring the regulation and supervision over health maintenance organizations from the department of health to the insurance commission, directing the implementation thereof and for other purposes*, 2015 https://www.officialgazette.gov.ph/2015/11/12/executive-order-no-192-s-2015/.

Republic of the Philippines, Congress of the Philippines. Republic of the Philippines, *Republic Act 11223: An Act instituting universal health care for all Filipinos, prescribing reforms in the health care system, funds thereof.* 2019.

Republic of the Philippines, Congress of the Philippines. *Republic Act No. 11469 An act declaring the existence of a national emergency arising from the Coronavirus disease 2019 (COVID-19) situation and a national policy in connection therein and authorizing the President of the Republic of the Philippines for a limited period and subject to restrictions to exercise powers necessary and proper to carry out the declared national policy and for other purposes. March 2020.* https://legacy.senate.gov.ph/Bayanihan-to-Heal-as-One-Act-RA-11469.pdf.

Republic of the Philippines. 2020. *Audit Report of the Philippine Health Insurance Corporation.* Commission of Audit. p. 7.

Republic of the Philippines. 2022. *Senate Bill 2505: An act creating the Philippine Center for disease control and prevention, 2022.*

Senate of the Philippines. 2019. Villar Files Bill Enabling Small Farmers to Recover From Typhoon Damage. Senate Bill 140 or the Free Index-Based Agriculture Insurance Act. Press release. 11 July.

Simeon, Louise Maureen. 2018. No More Farm Subsidies by 2021–DA. *PhilStar Global.* 27 May. www.philstar.com.

Social Security System (SSS). 2015. *Moving Towards Inclusive Growth: the Philippine Social Security System.* https://www.sss.gov.ph/sss/DownloadContent?fileName=2015_Updated_ISSA_Report_on_ISCoverage_FINAL.pdf.

———. 2021. The Social Security System granted ₱869.99 million worth of unemployment benefits to 67,937 members for the period of January to September 2021. Press release. 18 November 2021. https://www.sss.gov.ph/sss/appmanager/pages.jsp?page=PR2021_054.

The Geneva Association. 2021. *Public–Private Solutions to Pandemic Risk: Opportunities, Challenges and Trade-offs.* https://www.unisdr.org/preventionweb/files/77863_publicprivatesolutionstopandemicris.pdf.

Unified Accounts Code Structure. https://www.uacs.gov.ph/resources/faqs.

R. Wehrhahn. 2010. Insurance Underutilization in Emerging Economies: Causes and Barriers. In C. Kempler, M. Flamée, C. Yang, and P. Windels (eds.). *The Global Perspectives on Insurance Today: A Look at National Interest versus Globalization.* Palgrave Macmillan.

World Bank. 2020. The Philippines Parametric Catastrophe Risk Insurance Program Pilot. https://openknowledge.worldbank.org/bitstream/handle/10986/36013/The-Philippines-Parametric-Catastrophe-Risk-Insurance-Program-Pilot-Lessons-Learned.pdf?sequence=1&isAllowed=y.

World Bank. 2021. Philippines: Mandanas Ruling Provides Opportunities for Improving Delivery Through Enhanced Decentralization. Press release. 10 June. https://www.worldbank.org/en/news/press-release/2021/06/10/philippines-mandanas-ruling-provides-opportunities-for-improving-service-delivery-through-enhanced-decentralization.

World Health Organization (WHO). 2016. International Health Regulations (2005) Third Edition.

———. 2018. *Joint External Evaluation of IHR Core capacities of the Republic of the Philippines, Mission Report.*

———. 2018. Asia Pacific Observatory on Health System Policies. Health Systems in Transition. *The Philippines Health System Review.* 8 (2). 9789290226734-eng (3).pdf p. xviii.

WorldRiskReport. 2020. *WorldRiskReport 2020.* https://reliefweb.int/sites/reliefweb.int/files/resources/WorldRiskReport-2020.pdf.